**ACPL ITEM
DISCARDED**

MAY 0 5 2010

Whose Fair?

WHOSE FAIR?
Experience, Memory, and the History of the Great St. Louis Exposition

JAMES GILBERT

THE UNIVERSITY OF CHICAGO PRESS   CHICAGO AND LONDON

James Gilbert is professor of history at the University of Maryland. He is the author of ten books, including *Perfect Cities: Chicago's Utopias of 1893*, *Redeeming Culture: American Religion in an Age of Science*, and *Men in the Middle: Searching for Masculinity in the 1950s*, all published by the University of Chicago Press.

The University of Chicago Press, Chicago 60637
The University of Chicago Press, Ltd., London
© 2009 by The University of Chicago
All rights reserved. Published 2009
Printed in the United States of America

16 15 14 13 12 11 10 09   1 2 3 4 5

ISBN-13: 978-0-226-29310-3 (cloth)
ISBN-10: 0-226-29310-6 (cloth)

Library of Congress Cataloging-in-Publication Data

Gilbert, James Burkhart.
Whose fair? : experience, memory, and the history of the great
St. Louis Exposition / James Gilbert
p. cm.
Includes bibliograpical references and index.
ISBN-13: 978-0-226-29310-3 (cloth : alk. paper)
ISBN-10: 0-226-29310-6 (cloth : alk. paper)
1. Louisana Purchase Exposition (1904 : Saint Louis, Mo.)—
History. 2. Exhibitions—Social aspects—Missouri—Saint Louis.
I. Title.
T860.B1G553 2009
907.4′77866—dc22
2009003105

∞ The paper used in this publication meets the minimum requirements of the American National Standard for Information Sciences—Permanence of Paper for Printed Library Materials, ANSI Z39.48–1992.

CONTENTS

List of Illustrations · *vii* ·   Preface · *ix* ·
Introduction · 1 ·

ONE
Fair Itineraries: Experience, Memory, and
the History of the Fair
· 13 ·

TWO
Making History
· 37 ·

THREE
Making Memories
· 69 ·

FOUR
Making Images
· 102 ·

FIVE
Mrs. Wilkins Dances
· 123 ·

SIX
The Beholder's Eye: Making Experience
· 153 ·

SEVEN
Making Identities
· 185 ·

Notes · *195* ·   Index · *217* ·

ILLUSTRATIONS

1. Palace of Electricity and the Palace of Varied Industries with diminutive figures · 107 ·
2. Jessie Tarbox Beals · 110 ·
3. "Negritos Shooting Bow and Arrows" · 112 ·
4. "Triumphant Host of King Jesus" · 114 ·
5. Bird's-eye view of a parade · 121 ·
6. Photographing the nonwhite, type 1 · 125 ·
7. Type 2 · 126 ·
8. Type 3 · 127 ·
9. "Mrs. Wilkins, Teaching an Igorrote-Boy the Cake Walk" · 128 ·
10. Mrs. Wilkins sings · 129 ·
11. "The Extremes Meet—Civilized and Savage Watching Life Savers' Exhibition, Igorote Family at World's Fair" · 133 ·
12. Dr. Hunt of the St. Louis Pound serves stewed dog to David Francis while Igorots, portrayed as black Africans, look on · 140 ·
13. Poster advertising "Uncle Tom's Cabin" · 147 ·
14. Contemporary white version of the cakewalk · 148 ·
15. Judy Garland does the "cakewalk" · 150 ·
16. "Two Men Walking on a Road through the Tyrolean Alps" · 162 ·
17. "In the Irish Village—Pretty Lassies in Jaunting Cars before 'Cormac Chapel'" · 163 ·
18. "Crowd in the East End of the Pike" · 168 ·
19. The entrance to Creation · 170 ·
20. "Palace of Machinery Interior" · 177 ·
21. "Missouri Tomatoes" · 178 ·

PREFACE

In writing this book I have accumulated many debts that I can repay only with acknowledgement and thanks. I greatly appreciate the research support from the University of Maryland, which made frequent trips to St. Louis possible. My colleagues in the History Department have patiently listened to my ideas and offered useful suggestions during a preliminary presentation of a chapter at the department's Miller Center for Historical Studies. I am grateful to Jonathan Auerbach for his perceptive critique. I especially thank Saverio Giovacchini for an insightful and critical evaluation of an early version of the manuscript. My student, Erik Christiansen, both read and critiqued the manuscript and helped me secure the photographs that appear in the book.

Several years ago, in 2004, even before I decided to embark on this study, I was invited by the Missouri Historical Society and the University of Missouri–St. Louis to give the James Neal Primm Lecture marking the hundredth anniversary of the St. Louis World's Fair. This was my introduction to the historical society and the first of several trips to St. Louis and to that distinguished research center. I am enormously indebted to the staff of the society, Christopher Gordon, Sharon Smith, and especially Duane Sneddecker and Dennis Northcott for their help. The manuscript and photographic resources of the society, their accessibility and range constitute a model archive. I was also delighted to have the chance to know the city of St. Louis better, a sometimes unknown treasure for a Chicagoan like myself. I was particularly happy to introduce the city to my colleague Mark Micale, with whom I had several fruitful conversations about the project.

For any researcher, the Library of Congress is an unmatched resource. I used both their manuscript collections as well as their marvelous holdings of prints and stereographs that depicted the Fair. Their Fair materials in the general collection were surprisingly extensive. I also benefited from several months' tenure at the Library's John W. Kluge Center, where I also gave an early talk on my project. I am also grateful to the Schlesinger Library at the Radcliffe Institute at Harvard University in Cambridge for help in securing the work of Jessie Tarbox Beals.

I also had the good fortune to spend several months over two summers in Germany at the University of Erfurt. I presented a portion of this work to colleagues and students at a conference there in 2007 and then held a short course on writing the history of the Fair over the summer of 2008. I especially wish to thank my colleague Juergen Martschukat at Erfurt for his hospitality and interest in this project, and the students in my classes for their helpful suggestions.

I give special thanks to Tim McGovern and Doug Mitchell at the University of Chicago Press. I have worked on several projects with Doug, each one with a deepening admiration for his intelligence, encouragement, and learning. The readers he secured for the manuscript were both critical and smart, and I have tried in every respect to take their advice seriously. My final acknowledgement is to Chip Rothschild, who read, criticized, and then reread portions of the manuscript with a sharp and objective eye.

The title of this book represents my effort to capture the dual purpose of the work, which is to write about the Fair itself and, at the same time, step back and consider what that task is really about. To whom the Fair belongs is an issue that thrusts us to the center of this consideration. Is it the historians who examine the documents and archival materials, who construct narratives that interpret the event for contemporary eyes? Is it those who remember the Fair in some fashion as part of the paternity of contemporary St. Louis, as a heritage and that city's usable past? Is it the visitors to the Fair, whose experiences have vanished with the passing of several generations, but whose reactions are suggested in memoirs and oral histories and the traces of their meandering around the fairgrounds? Or is it the possession of all of us, those from the past joined to our own time? Can we not imagine ourselves a community of wanderers and dreamers among its crowded display halls and fanciful concessions trying to make sense of what we observe?

INTRODUCTION

The enduring difference between history and memory is a major challenge to our understanding of the past. It is a quarrel that is immensely complicated when one asks what either narrative has to do with actual, lived experience. Historians rightly revise memory by summoning up a larger perspective, deriving their analysis from hindsight. On the other hand, the custodians and keepers of memory appear far more able to communicate their accounts widely. They commonly command more public attention; their perspectives often determine public commemorations and dominate other forms of popular remembrance culture. Consequently, memory, and particularly, collective memory, in all its guises, is probably what most Americans understand as history. Brimming with political and emotional energy, memory is tenacious and persistent. As with any powerful narrative, collective memory is also selective memory. From its perspective, historians seem to focus on aspects of the past that are largely invisible or irrelevant, surprising, and even disturbing. On the other hand, historians appear less than enthusiastic to embrace the stories that collective memory most cherishes; they are less concerned with objects of veneration, historic sites, monuments, revered texts, and symbols. If memory and history derive from the same events, why do their accounts radically diverge over time? Is it because historical thinking is unnatural and impersonal, striving always for the unachievable goal of objectivity? Or because memory is self-interested and deliberately subjective, focusing on the limited perspectives of the individual or the emotional validations of community and national self-justifications? Or is it because the two approaches represent different and profitable ways to consider the same past?[1]

This book confronts head-on the competition between accounts of the past that we all share. Even historians in their everyday lives (were they to admit it) quite readily participate in affirming collective or individual memories of one sort or another that they might question as professionals. And just as history can include various definitions, perspectives, and emphases, so memory consists of individual recollections that place the teller at the center of events as well as collective memory, which consists of the shared stories that constitute cultural proficiency.

I begin with the bias of a historian. But I also have a deepening recognition that our changing and restless accounts of historical events—the ever-shifting if always fascinating unfolding of new interpretations—is adding to the complexity and the instability of historiography, if not of history itself. Having watched the profession ascend one position after another in attempting to gain a superior perspective on the past—first political history, then economic history via Marxism, then "history from the bottom up," then economic and political metrics and modernization theory, then cultural history, then gender history, then theory and more theory that looks back and deconstructs these interpretations, and most recently the neonarrative and international history—one could wonder whether we are building higher as we enthusiastically embrace one construction of the past and then another. In defense of this dialectical development, I know that the past always looks (and perhaps becomes) different as the present pushes forward. Inexorably, our understanding of the past opens backward in tandem with the forward rush of time. Thus past and present thrive as reflections and dependencies of each other. As Harry Harootunian puts it, when history strives to be different from memory, it "must always be a history of the present, which means a politically driven history."[2]

Similarly, it is impossible to overlook the problems raised by memory, its self-absorption, subjectivism, unreliability, and false certainties. These are compounded by the pragmatic aims of collective memory. Sometimes these aims are overtly political in nature, or chauvinistic, or sometimes merely narrow, nostalgic, or focused locally. They can amount to little more than raw boosterism. Their origins in oral history and their personal agendas can be deeply troubling. Collective memories often deny the brute failures of life, the tough version of reality, shaping a narrative of happy endings and American triumphalism. At the same time, they often seem closer to a sort of democratic and accessible history than specialized monographs fashioned for academic publications do. Sometimes I suspect that we historians dismiss memories, whatever their character or origin, simply because they seem trivial or idiosyncratic. Yet memory, unlike academic history (sometimes), represents a genuinely usable and used past. One of its characteristics, for good or ill, is its apparent fixedness, its slow, hesitant change, and its recollection of a past that few wish to challenge. Historians invented a word to describe their ever-evolving conceptions of the past, hence the term *historiography*. But the study of memory has never devised such a word. We can only affix some adjective like *changing* to memory to describe any alterations we perceive. Such changes are apparently so infrequent that no special vocabulary yet exists to indicate its evolution. Yet one of the most fascinating, if generally less noted, aspects of memory is the initial rapid transformation of individual experience into a narrative that fits into the larger format of a developing collective memory.

My dissatisfaction, uneasiness, and fascination with these competing narrative forms has become even stronger because I fear that historians as well as memory-keepers sometimes neglect the experience of certain types of historical actors, particularly those who constitute the audience. This is a serious allegation, but it seems obvious that we historians often discount historical experience (and memory) in lieu of grander formulas of explanation derived from our concentration on the record of elite actors and opinion. As David Lowenthal suggests, what is historic and remembered is only a part of what we do or live.[3] Of course, our neglect of the missing "others" may only be the inevitable result of time and distance. We cannot ever confidently know what people did or saw or thought about what they were doing or seeing or thinking. So we sometimes just ignore the problem. Certainly the limited perspective of the historical moment provides little elevation from which to gain perspective and understanding. It may simply be impossible for participants to grasp the meaning of a historical event without the passage of enough time, until, as it were, all of the implications have marched into view. So too memory, even before it becomes more or less fixed, may be radically different from experience because of the changes wrought by conversation, reading, and the interpolation of collective experience into individual impressions.

Nonetheless, in this book I propose to pay attention to experience despite its obscurities and mystery, to see what a careful rereading of documents can reveal about one of the great, important events of the early twentieth century: the Louisiana Purchase Exposition held in St. Louis in 1904.* So this is a

---

*The fair was officially titled the Louisiana Purchase Exposition, but it went by many other names in the press, guidebooks, and conversations of its officials and visitors. These include the St. Louis World's Fair, the St. Louis Exposition, and the Louisiana Purchase Centennial Exposition. I will keep these designations in quotes and sometimes employ them myself.

book about the Fair but at the same time not limited to the Fair, because it inquires into how we come to know about and understand such great public events and what we should make of them. It is an essay about methodology and the history of an event that uses both categories to enlighten and structure the other. I intend, then, to read through historical accounts and the sources upon which they are based and to explore memory sites and celebrations and their sources to discover the actual experience of fairgoers insofar as I can reconstruct it. I propose to use my findings to suggest alternative ways of understanding this historical event, both as written history and as memory. The result is, I think, a different story from what we think we know. At the same time, I hope to offer insights into ways that history and memory might complement each other, learn from each other's proceedings and accepted methods and truths. Most of all I hope to restore a respect for the actors of the past and their making of that past. No doubt they were trapped in worlds of assumptions and unarticulated causes not of their making; they were players on a stage constructed for them by prior events and handed-down truths and by events the significance of which they were only vaguely aware of even as they lived through them. They handled props designed elsewhere and lived in a universe of discourse and representations invented by others. Yet they spoke their lines unrehearsed. They often missed cues meant to guide them to certain conclusions because of private musings, idiosyncratic desires, and unconstrained imaginations. They frequently did not see what was intended to be seen, or they misread it, sometimes willfully. It is the sum of this marginal behavior that I wish to restore to the center of the historical account. In this quest, experience will be my constant, if elusive companion, guide, instructor, and critic.

But why choose St. Louis as the site to explore such issues? I do so principally because of the well-established historical narrative that has developed to explain the meaning of the Fair based upon the works of Robert Rydell then adopted and expanded by many others. This interpretation places matters of race and imperialism at the center of the intention of the Fair builders and sponsors, and it argues that this is the predominant meaning of the Fair. Because of its power and persuasiveness, this reading has become widely accepted. Just as rich is the unusual memory culture that has grown up around the Fair and its role in the identity of St. Louis. Even today, this event maintains an enduring presence and power, representing both the high point of a rich urban history as well as possibilities of a future that never materialized.

My interest in this subject began several years ago while I was researching a book on the World's Columbian Exposition of 1893. That splendid and most-written-about fair is also the subject of considerable hyperbole, much of it inspired by the fair administration and city politicians in their desire to set Chicago and its celebration of Columbus as the crown jewel in the American 1890s. The greatest fair of the century! The city of the future! The most magnificent exposition in history! The most widely attended fair until the modern period! It was this last assertion that gave me pause. The fair administration claimed that twenty-seven million persons attended the Columbian Exposition—or at least they implied this by failing to disaggregate the figure, thereby allowing others to reach unwarranted conclusions. With the entire population of the United States at that date around sixty-seven million, this figure implied a rendezvous of extraordinary proportions, a pilgrimage to surpass almost any mass movement in human history as almost 40 percent of the entire nation gathered by Lake Michigan to celebrate Chicago's festival. Except that it didn't happen. And historians who have uncritically repeated this figure have only perpetuated one of the myths about world's fairs in general that cry out for correction.

A similar extravagance infects memories of the fair held ten years later in St. Louis, although in this case it is principally emotional exaggeration rather than numerical inflation at stake. Frequently, if not always, the Fair is remembered today in the context of the 1944 film *Meet Me in St. Louis*, with Judy Garland and a host of wonderful character actors, including Margaret O'Brien. This Hollywood version of an already nostalgic memoir written by Sally Benson and published in the late 1930s in the *New Yorker Magazine* portrays the lives of an upper-class family over the year leading up to the Fair. Benson, born in St. Louis and a successful film reviewer and later screenwriter, also wrote *Junior Miss*, a series of short stories made into a 1945 film. The four "acts" of *Meet Me in St. Louis* are set apart visually by the insertion of sentimental drawings of the four seasons that come alive—a likely sign in Hollywood symbolism of the happy past. The story recounts the opportunity of the father to move to New York for business reasons. He is resisted by his daughters, who are engaged in budding romances with neighborhood boys. It is also the last year in town for several boys before they head off to Princeton.

But the Fair, abounding in splendor and promise, eventually convinces the father (with some aggressive encouragement and emotional blackmail by his daughters) that St. Louis is worth it after all. The movie ends with romance triumphant. But this is, after all, 1944. And when Judy Garland plaintively sings "The Boy Next Door," I doubt if the thoughts of the audience were on trains pulling into Princeton Junction. More likely their attention was to troop ships landing in Normandy or Okinawa. Thus, nostalgia for this special past is intensified by fears of the moment—another example of Hollywood's iconic poetic mode, creating a stereography of emotion in which the viewer experiences both past and present as part of the same dimension. The movie itself re-creates a past that is the more treasured because of the toll of war, the dislocations of modern social and cultural change, the decline of small towns (and cities), and, more broadly, the disappearance of the Victorian American heritage. As a feature intended for all of American society, not just Missouri, the film bathes the early nineteenth century in the soft light of a simpler, family-oriented time.

To view this film as part of a remembrance of the Fair, then, is to push memory through the sieve of wartime emotion and then mix it with a variety of recollections, some lived and some imagined. For young people living in St. Louis in 1944, it accorded with Hollywood's version of American history, the stardom of Garland, and celebration of an old-fashioned world that was a frequent subject of movies at the time. But the Fair is scarcely the principal issue. And rather than the vision of the future it was intended to be in 1904, the Louisiana Purchase Exposition had become an expression of Hollywood past perfect. This reversal of meanings may have been anticipated by certain nostalgic and historic elements already present in the 1904 Fair itself. It had optimistically anticipated a future that would shortly be cut down on the killing fields of the Marne and Verdun. In the collective memory of later generations, the Fair represents something quite different from what builders of the Fair intended. It signified a moment marking the decline of St. Louis and expressed a growing feeling of uneasiness about the urban future, not its celebration. It represented a high point of a middle-class Victorian civilization that was about to vanish and that exists now only as a point of reference for less optimistic generations that followed. A city of over 580,000 souls in 1904, St. Louis retains only about 60 percent of that population today.

Inevitably, a book about history, memory, and experience suggests serious theoretical issues that I shall raise when they become relevant. These are issues of considerable complexity and importance that go straight to the heart of recounting the past. To explore the interaction of history, memory, and experience immediately suggests the tangle of theoretical problems that have been winding for several decades around such tragic events as the Holocaust and World War II. Clearly the St. Louis World's Fair is not a catastrophic event like these. And yet the same problems appear. What do we mean by *memory*? Whose memory counts? Which has priority, collective or individual memory? Aren't all memories shaped by the imposition of outside narratives and by cooptation by forms of commercial cultural retrospectives like films, exhibits, newspaper accounts, and novels? *Experience* is an even more problematic term for two reasons. First, it is difficult, if not impossible, to identify individual experience that is unshaped and uncontaminated by some instrument of collective memory. Even if we could identify individual experiences, is there a practical means to sort and combine these into useful historical categories—to make them sources rather than illustrations? And second, we must ask, does experience really matter to historians? What if we decide that the historical meaning of an event exists independent of what the majority of participants might have experienced and remembered? Does this confirm the adage that the historian is someone who tells you that your memories are wrong? My purpose in this book is to look through the writings of historians and the activities of the custodians of memory to identify and reconstruct experience, to explore what can be learned from a closer look at the visitors to the Fair. I intend then to reintegrate this experience to see if this source changes our historical understanding.

To suggest that we can easily discover unadulterated experience, the raw reaction of fairgoers in St. Louis in 1904, is not my intention. Experience is inseparable from culture, and it is always susceptible to modification once it is recalled, recounted, and discussed. Passing into memory, it takes on two forms, individual and collective. So experience is remembered as what is individually important and also as part of a larger, collective rendering. There can be considerable dissonance between these two forms, when individual memories do not match a developing master narrative of an event; nonetheless, they almost always interact. Sometimes, personal memories absorb elements of collective memory that the individual did not experience. As for its transformation into history, experience is often assumed or subsumed into broad generalizations without much attention to the serious theoretical problems it provokes.

If experience is an ephemeral event, a moment in a stream of meanings, what is to be gained by thinking we can isolate and explore it? By *experience,* I mean two things: behavior, which we can record and measure and interpret, and comprehension and awareness, whose significance we may only surmise. Nonetheless, the attempt to reconstruct these two elements of experience has important consequences. First, it demands that we ask different sorts of questions about the meaning of events and their participants. Furthermore, it forces the historian to look at sources differently and to consider some sources that are underused. Finally, it forces us to pay attention to reception and to estimate what audiences might have understood and what ideas, prejudices, and notions they might have deployed as part of this understanding. I have not discovered some unknown source, some unexplored cache of documents that will suddenly reveal how the audiences engaged the Fair. Instead, I propose to read the voluminous record of the Fair guided by a different set of priorities.

There is a brief but vivid way to illustrate the pitfalls of ignoring the circumstances of experience. At the dedication day ceremony in St. Louis on April 30, 1903, President Theodore Roosevelt gave a speech that has become a document widely used by historians to determine the meaning of the event and its place in the history of the time. But there is a problem. Among the thousands who thronged to the event, the French ambassador was privileged to sit on the dais with the president. A shame it was, he reported later, that hardly anyone in the auditorium could hear the speech because of poor acoustics in the hall and the prevailing crowd noise. Did Roosevelt's words thus die for lack of immediate auditors? Are they important because they were subsequently reproduced in print? If this is the case, do they have anything to do with the Fair, or was Roosevelt just using the occasion as his pulpit for political ideas expressed elsewhere? And what did the event mean to the audience?

To approach this problem, we need to think more generally about living in a world of unamplified sound, a world of "megaphone men and criers." As one writer described the amusement portion of the Fair, the clamor of shouts and shills began at the St. Louis train station and extended all the way to the entrance to the Pike in a cacophony of "facial blunderbusses."[4] Or what about circumstances where the audience could not hear? How many of those present at other such commemorative events during the summer of 1904, especially those delivered outdoors, actually heard the grandiloquent speeches about the purpose of the Fair? How much attention should we pay to such speeches in a world without loudspeakers or electronic reproduction? How should historians assess the difference between words spoken before an audience and those reproduced later for a reading public? What did it mean to attend without hearing? How did the organizers of public events, who were obviously well aware of the problem, shape performances and speeches for an era before the electronic reproduction of sound? Does this explain the popularity of parades, with their blaring brass bands and pantomimes? As I shall discuss, there were a variety of interesting steps Fair organizers took to convey meanings in this world of auditory limitation.

This book is divided into seven chapters, the first six of which begin with the exploration of a theme and then move into consideration of theoretical issues that cluster around our methods of articulating that theme. Chapter 1 is devoted to placing the world's fair into its surrounding contemporary cultural contexts. It examines the function of fairs, especially those held within the United States during this period. Trying to isolate fairs from other similar events like amusement parks and trade shows exaggerates the importance and singularity of the Louisiana Purchase Exposition. It is also a distortion to depict it outside of a developing European/American culture of world's fairs. The chapter will move then to discuss nineteenth-century display and pageant culture, as well as tourism, all of which were represented at the time in St. Louis. Then it will place the St. Louis Fair in its specific historical context, emphasizing the special circumstances that brought it to the city. Moving from there, the chapter will end with an analysis of the intellectual history of the Fair as it was reflected in its important Congress of Arts and Sciences, which attracted hundreds of important thinkers from the United States, Europe, and beyond.

Chapter 2 begins with the voice of the Fair itself, as articulated by David Francis, the energetic and accomplished president of the Louisiana Purchase Exposition Company. Francis is important for several reasons. It was his vision that the Fair embodied, although, of course, much of that plan was derivative and predetermined by the developing European and American tradition of world's fairs. But Francis is also central because of his activities after the Fair ended in December 1904. He became the president of the first official memory organization devoted to preserving the cultural and economic effects and artifacts of the Fair. Perhaps most important, he shepherded the preservation of the archives of the Louisiana Purchase Exposition Company as well as his own extensive papers. These have ever since been the major source for historians seeking to write about the Exposition. And if these historians have by and large repeated his perspectives (if not always his values), the explanation rests in part on the richness of these sources.

While Francis established a record that allowed him, *in absentia*, to repeat his ideas through the writings of later historians, the Division of Press and Publicity of the Louisiana Purchase Exposition Company worked to create an immediate national perspective and impression that would energize visitors to travel to St. Louis and attempted to create expectations for what these visitors might encounter. The scope of their work was astounding in its extent and particularly effective because the national press and periodicals reproduced, verbatim sometimes, the releases that advertised the Fair. We learn much when we explore how this publicity portrayed the Fair, what it emphasized, and then what the national press reproduced from this material. This becomes doubly significant because, as will be seen, the audience actually present at St. Louis was far smaller than has been assumed. So, in the end, publicity may be the most significant influence in determining the contemporary meaning of the Fair in its national setting. Beyond publicity, there are two special forms of literature that depicted the Fair, and they too may have had considerable influence in shaping its contemporary meaning. These include guidebooks and official publications, as well as widely sold novels about the Fair that were apparently written using guidebooks as sources.

The final section of chapter 2 will explore the historiography of the Fair and what writers of our own day have made of the event. While the general interpretation is established—and there exists a master narrative of sorts—not all historians emphasize the same elements nor come to the exact same conclusions. However, generally dependent upon the archives, they reproduce many of the perspectives of Francis and his board, and indeed, they repeat the national political priorities expressed at the Fair by politicians. At the same time, this historical literature is deeply critical of the racial and gender assumptions of these creators. It rightly deplores the intense (even grotesque) anthropological scrutiny of colonial and dependent and conquered peoples on display in special enclaves and sideshows that characterized this and other late-Victorian fairs. And if it focuses on the imperial aspirations of a nascent American empire in the publicity bravado and bully speeches of celebrities and politicians, this is because these were the apparent purposes of the Fair.

Chapter 3 also begins with the voice of David Francis, but in this case, wearing his other hat as leader of the first official memory organization generated by the Fair. In this guise, Francis not only worked to preserve documentation of the Louisiana Purchase Exposition, but he also sought, as all subsequent memory organizations would do, to create a collective memory for St. Louis and, to some degree, for the rest of the country. Rather quickly this became largely a St. Louis enterprise, with the local impact its chief focus. Almost immediately, then, memory, as reflected in the Louisiana Purchase Historical Association organized by Francis, became more localized and part of the city's economic and cultural heritage. This memory also became, increasingly, nostalgic for better times as it became obvious that the city had lost out in its competition with Chicago. Because of the efforts of Francis and other city leaders, St. Louis established a lively memory culture, with frequent celebrations, reenactments, and displays of items from the Fair. After *Meet Me in St. Louis* was released in 1944, this film became a principal, if not the main, vehicle for recollecting the Fair and was regularly shown at commemorative events. As time passed, the custodians of memory became persons who were too young to have attended the Fair, until today, the leading memory organization is constituted entirely of later generations of St. Louis–area members.

The sources of memory, if not collective memory itself, are not, however, simply confined to organizations, but constitute important historical documentation existing in other formats and gathered at different times. These survive in two general forms: contemporary published personal accounts, diaries, and unpublished accounts and oral histories collected much after the fact (in 1979, during the seventy-fifth anniversary of the Fair, for example). This chapter will explore some of the theoretical and practical problems raised by collective memory that are revealed by memory organizations and official commemorations, as well as the perils and promises of using individual oral history and diary accounts.

Chapter 4 focuses on the photographic record of the Fair. This is available in a variety of formats: photographic postcards, official photographs, shots taken by professional photographers, and, in particular, the scarcely used but huge stereographic record. I will argue that these different formats depict very different impressions, experiences, and purposes. They seem to show several different Fairs, rather than a singular, coherent vision. I will concentrate in particular in this section on the struggle of photographers to understand and depict the extensive anthropological displays of peoples and cultures.

Chapter 5 suggests how we might think about photography in its multiple contexts. I focus on one photograph, a remarkable shot taken by Jessie Tarbox Beals entitled "Mrs. Wilkins, Teaching an Igorrote-Boy the Cake Walk." This astounding depiction of a white woman dancing with a dark, almost naked Filipino youth, suggests the need to inquire closely into ways that the visitors might have comprehended the omnipresent enactments of racial sentiment at the Fair. This photograph violates the carefully prescribed anthropological ideas and boundaries established to contain and explain race in the modern world and specifically in the exhibits of the Fair. It also challenges us to re-think generalizations about race made by contemporary historians. While it is unusual, it offers heightened possibilities for demonstrating what a deep reading can accomplish. To understand this photograph, I will explore the varied historical contexts within which it might have been viewed at the time, linking it, as well, to other transgressions of racial codes that exploded on the fairgrounds. This discussion will include an examination of the overtones of American racial thinking current at the time as they were reflected in and challenged by Mrs. Wilkins's dance.

Chapter 6 will consider the larger question of individual and group experience. Reading through the historical and memory sources available and using the photographic evidence, it becomes possible to reconstruct, in some detail, how patrons at the Fair spent their time and money. This allows us to speculate about how individual visitors might have understood the event and, even more important, how they may have imagined themselves interacting with the people, things, and ideas on display. Beyond revealing which displays and concessions were most popular, this chapter will explore the function of state and foreign nation exhibits, the particular attraction of several concessions on the Pike, and then how special groups visited the Fair. Because hundreds of national organizations made St. Louis their convention headquarters for 1904, probably more than 100,000 visitors attended under the auspices of some organization. Others came to watch the Olympic Games held for the first time in the United States that year in St. Louis. Some came for the intellectual congresses. And, significantly, many were attracted to family reunions held on the fairgrounds.

My study will conclude with a short essay on conclusions and generalizations. In chapter 7, I will suggest several new ways to construe the meaning of the St. Louis Fair and world's fairs in general—ways of reflecting upon history with the new knowledge gained from considering memory and experience as vital sources of meaning. To discover those meanings is the purpose of this book.

CHAPTER ONE

FAIR ITINERARIES
Experience, Memory, and the History of the Fair

The great world's fairs at the turn of the nineteenth century are among the most extravagant cultural events staged in modern history. Highlighting the peak years of imperial expansion when Great Britain, Germany, France, and the United States jostled to attach the remaining "uncivilized" territories of the world to their booming industrial economies, they were extravaganzas of optimism, showcases of the present, and predictions of the future, as well as peaceful environments for competition and display of goods and ideas in the edgy years of the early twentieth century. Like the contemporary Utopias invented by novelists and social theorists, they were also curiously trapped and defined by the past, simulacra of a world that could only imagine perfection as preserving the better half of the present and discarding or denying the rest. To say that they represented a secular equivalent of Lourdes or Mecca is only to underscore the almost sacred rhetoric invoked to describe them.

For historians, they have proved to be irresistible sources for understanding the transition to the modern world of consumerism, mass culture, and urbanism. If these idealized environments were transitory, like imaginary islands constructed amidst waves of turmoil and the daily struggles of industrial society, they also embodied the profound illusions of their day. They were visited by millions of tourists, almost all of whom traveled to these ephemeral destinations by one or another form of railway conveyance, converging on the host cities along the great main lines and, once there, circulating on local rails and tramways. They were isolated parks, simulated cities that, in fact, excluded much of the real city that hosted them. Although the 1930s represent a similar high point in the history of world's fairs, albeit with somewhat different purposes and challenges, the late-Victorian world's expositions can be seen as even more important in broadly illustrating and engaging the civilization of their day. Among these, the fairs at Chicago (1893) and St. Louis (1904) were rivaled only by the Exposition Universelle in Paris (1900).

Quite rightly, historians have carefully surveyed these temporary cities of cultural and social display, and they have generated a sometimes contentious, but often enlightening, historiography. Yet much remains to be done. Curiously, as more writing appears to explore these fairs and assess the culture they seem to reflect, the problems of interpretation intensify. With the appearance of each monograph or museum exhibition, the clash between history and memory increases. The longer we look, the more the focus seems to fade and blur. For these good reasons, it is impossible to put off a discussion of how memory depicts these events and how both memory and history have pondered or ignored the actual experience of the fairgoers themselves. I intend to explore the difficult and elusive question of experience and then consider if and how attention to what people did and saw, how they enacted and recorded their experiences, might change our historical conclusions and amend collective memories of such cultural events. Increasingly it is apparent that there is a clash among history, memory, and experience, three categories representing three bodies of constituencies who have engaged these monumental events at different times and for different purposes.

In the beginning of analysis there is the problem of magnitude and shape. Were world's fairs the monumental events proclaimed in city booster literature? How does size measure their impact? Historians and the keepers of collective memory have been impressed, even astounded, by the attendance figures claimed at Chicago and St. Louis. Almost every history, encyclopedia entry,[1] and article, as well as the collective memory of both events, has repeated the same figures: twenty-seven million patrons in Chicago in 1893 and around twenty million in St. Louis nine years later. The Chicago figures, if true, would mean that over a third of the nation descended on that city during the summer of 1893 (during the height of a terrible economic depression) and that somewhere around 20 percent traveled to St. Louis. Yet even a cursory disaggregation of attendance reports of both fairs indicates that these are wild overestimates. The quoted figures are the raw records of one-time entries onto the fairgrounds; they do not necessarily count separate individuals. In the case of Chicago, about seven million of these entries consisted of the comings and goings of employees; about the same number is also true in St. Louis. Subtracting the number of employee turnstile entries, the

St. Louis Fair administration itself estimated that about twelve million entries represented actual patrons. And yet, this too must be judged a huge exaggeration, for it does not count the frequent trips that most citizens of St. Louis and out-of-town guests made. Even with the automobile civilization of the late 1930s, making short visits plausible, the administration of the New York World's Fair of 1939 determined that the average patron attended more than two times. In St. Louis, where travel and tourism were dependent upon trains and hotels and visits to relatives, the multiple of visits is likely to have been even more. And so, we are left with a good estimate of somewhere around four million different individuals who went to the fair in St. Louis—an impressive, but hardly remarkable, number.[2]

The Fair administration clearly knew that its published figures were exaggerated or at least susceptible to a misinterpretation that it nonetheless left standing. Its final internal report listed 12,804,616 paid admissions (indeed all of its internal documents quote the same figure) and over 7,000,000 free admissions. The latter number included visiting dignitaries and Fair officials but (mostly) workmen. There were also small numbers of free admissions on Sunday.[3] Not only did the administration understand this after the termination of the Fair in calculating profit and loss, it was well aware even in the planning stages that world's fair figures in general were likely to be misleading and overstated and puffed up by exaggeration.

An extended conversation in 1903 between Norris Gregg, chief of the St. Louis Concessions Administration, and Paul Blackmer, who had been charged with revenue collections at the Chicago World's Fair of 1893, explored exactly this point. It was Blackmer's contention that money earned on the Midway in Chicago (and the proposed Pike at the St. Louis Exposition) determined the financial success of the Chicago fair and the possible solvency of the upcoming St. Louis extravaganza. He had much to say about how to deal with the various concession companies that were making the rounds of world's fairs exhibiting exotic displays like the "Street in Cairo." Blackmer's prejudice against "Oriental" concessionaries was vivid, and he warned Gregg about problems with both specific owners and more generally with certain nationalities and their proclivity for theft.[4] In self-serving fashion, Blackmer claimed that the attraction of concessions and entertainments at the Chicago fair accounted for about 30 to 40 percent of all admissions. Furthermore, he estimated that of the twenty million paid admissions (less the workers and free entries) about fifteen million came from Chicago. In other words, Chicagoans (based upon the population of a million souls) went multiple times. Further, he estimated that only about five to six million admissions came from outside Chicago. It is worth reproducing the exchange that followed:

GREGG: So you think not over 5 or 6 million of the people who paid admissions at Chicago came from outside of Chicago?
BLACKMER: That is my belief.
GREGG: Some of those people who came from outside of Chicago would attend several times, would they not?
BLACKMER: All of them came two or three times.
GREGG: That is you infer that not over two or three million people came to Chicago to attend the fair?
BLACKMER: I believe that, and you would too if you came down there and saw the empty hotels with no people in them.[5]

If one adds the million or so inhabitants of Chicago (the unlikely possibility that every man, woman, and child attended) to this number, then Blackmer's estimate of different patrons falls to less than five million. Although Blackmer's figures seem inconsistent and extremely low, it is entirely appropriate, I believe, to extrapolate this reasoning to St. Louis, where we can estimate somewhat fewer different patrons, perhaps around four million.[6]

It should be clear why world's fair promoters might release figures that could easily be misconstrued to suggest that each admission represented a separate individual. The Division of Press and Publicity and the national press happily obliged by passing on this deceit. Huge numbers, in themselves, were deemed to be good advertising, and, anyway, a paid admission was a paid admission. But for historians, the reason may either be a desire to maximize the importance of the event for the telling of a good story or, more probably, just inattention. In any case, a more careful accounting of attendance fundamentally alters the way we should estimate the importance of these events. We also need to be doubly cautious that, in our enthusiasm for vast audiences, we do not see them as a mute block of observers, ready and willing to accept and absorb the various messages and lessons intended by the designers of the Fair.

There is another anomaly of attendance that should be mentioned here, if only to indicate something about the character of the audience—something which is obvious from photographs but generally has gone unremarked. That is the relative scarcity of children. While there are hundreds of photographs depicting adults in every portion, corner, and activity of the Fair, only a few contain children, and most of these are shots taken along the Pike. Admission figures confirm this visual impression that the Fair was primarily an adult event—and self-consciously aimed at adults. The "Final Report of the Department of Admissions" determined that, in terms of paid admissions, adults outnumbered children by about twenty to one. Figuring in free passes to schoolchildren, the ratio is still around ten to one.[7] This age distortion reflects, by the way, what Walt Disney discovered when he opened Disneyland in the late 1950s. For the first few years or so of his cartoon fantasy version of the world's fair, adult patrons outnumbered children four to one.[8]

The Louisiana Purchase Centennial Exposition was an immense and exhilarating experience for the city of St. Louis; in terms of acreage, it was the largest world's fair ever held. But size and space aside, this was an event within the crowded history of other world's fairs and expositions that appeared during the years of the late nineteenth century and early twentieth. Walter Benjamin, reflecting on Paris, which held several major expositions in the late nineteenth century, declared, "World exhibitions are the sites of pilgrimages to the commodity fetish."[9] For Benjamin, the magical power of things, materialized in the endless cascade of products of the Industrial Revolution, drew millions to the sites of their display. As Neil Harris aptly put it, world's fairs "performed as sites for self-discovery, camp meetings for a dominating middle class."[10] Before St. Louis, there were fifteen or so large international expositions.[11] But this is hardly an exhaustive list of the major fairs and exhibitions held during this period. Each decade from 1880 through 1910 witnessed more than thirty such events in Europe and the United States that attracted millions of patrons for a progress report on the rapid expansion of consumer culture. At these venues, these patrons saw many of the same exhibits featured at world's fairs.

But this general late-century exposition culture was not the same everywhere. All of the major events were held in European or American cities. But almost every such European fair was housed in a capital city: Paris, London, Vienna, and so forth. In the United States, this has never been the case. Washington, D.C., has never hosted a world's fair, although one could argue that the Capitol Hill Mall museums and appended shopping streets contain many features of a permanent world's fair. In fact, the early pattern of American fairs records something different from the centralizing culture of a Paris exposition. Instead, there was a distinct westward march through time and space: Philadelphia first, then Buffalo, Chicago, St. Louis, Omaha, and San Francisco. Fairs often commemorated important historical events, such as the French Revolution or Columbus's first voyage to the Americas. American events were also generally historical commemorations, but set at the edge of the advancing frontier and celebrating the physical expansion of the United States. American fairs left a path of cultural footprints across the newly claimed, conquered, and settled continent. Frequently this was accompanied by a specific commentary about Native Americans and their inclusion in the event, not as the (recently) dispossessed occupants of the land now turned into city and state, but as a curious, once proud, and now dying race.

As the Laird and Lee guidebook caption for the photograph of a statue at the Lewis and Clark Exposition of 1905 in Portland, Oregon, put it, "The various dances of the North American Indian have been famous in history and literature since the discovery of the country, and the passing of the Indian will be remembered for centuries by their quaint customs and usages."[12]

Frequently, the contemporary narrative of an American fair stressed the same rhetorical question decked out with self-serving innocence: How could a new nation, with its upstart cities thrown up in a building burst of only fifty years or so, pretend to host the world's greatest exhibitions? And yet, that is exactly the challenge they undertook. So the story of American fairs, at least until after World War I, almost always contained an astonished commentary on the dramatic industrial growth of the United States. There is an important truth buried here beneath these plaudits of self-congratulation. In 1904 the United States was still largely a rural and small-town society and would remain so for several decades. And the nation had managed to populate some of the largest and most dynamic cities in the world, the emporia of westward expansion. But an important question arises about this urban anomaly in the midst of a huge agricultural endeavor: Were American world's fairs primarily agricultural exhibits, just overgrown state agriculture fairs disguised as cities by the transitory plaster Beaux-Arts display halls, parks, and promenades? I think the answer is no, but I also believe that historians have greatly underestimated the importance of agriculture as the attraction and purpose of these American events.

St. Louis had, until recently, just such an agricultural fair that anticipated some of the elements of the 1904 Exposition. Founded in 1856 and closing in 1902, just before the Louisiana Purchase Exposition, the St. Louis Agricultural and Mechanical Fair fairgrounds occupied a large area complete with an amphitheater for horse racing, minstrel shows, theater, and other performances, and a small zoo. In some respects, this fair promoted St. Louis on a smaller scale than the later Exposition. Thus, in 1887 the newly elected mayor, David Francis, invited President Cleveland to the city to attend the fair as well as the Veiled Prophet pantomime. "Our fall festivities are at their height," at this time, he wrote. The streets of the city would be "brilliantly illuminated by arches of blazing light," and "our exposition and our agricultural and mechanical fair, each the greatest of its kind on the continent," would be in progress.[13]

There is something else to be noted about the extraordinary congregation of fairs and exhibitions at the end of the nineteenth century. Casting a wider net, considering not just events specifically designated as world's fairs by the Bureau International des Expositions but also the huge number of minor contemporary fairs, it becomes clear that world's fairs are primarily larger versions of what could be seen at many other venues under other auspices. There were trade fairs, art and industry fairs, agricultural expositions, colonial celebrations, historical commemorations and reenactments, and spectacular entertainments like Buffalo Bill's Wild West show. Added to this list must be the growing number of amusement parks, such as Coney Island, that reproduced the most popular features of fair midways.[14] World's fair culture needs to be seen as a specialized version of the era's new burst of urban tourism and leisure activity. As the circuit for commercial concessions like the Street in Cairo expanded, many of the displays at such venues as St. Louis could be, and were, seen and experienced elsewhere. As I will discuss, the St. Louis Exposition, like all other international fairs, was an eclectic collection, combining features from existing traveling exhibitions, amusement parks, trade shows, and public performances. In fact, the citizens of St. Louis were already familiar with many of the elements that constituted the 1904 Exposition. Even the vaunted nighttime lighting of buildings had been visible earlier, during Carnival Week in June 1899, when Anheuser Busch and Lemp breweries were made "brilliant at night by thousands of electric lights."[15] So the Fair was both new and not new, innovative, and yet part of a burgeoning culture of world's expositions and lesser commemorations where many of the same exhibits and concessions repeated their performances, where an expanding international group of specialized entrepreneurs moved from event to event. Thus, it became an important and necessary strategy of any fair administration to make what was continuous look discontinuous and revolutionary and to claim innovation when it was often merely repetition.

An even broader context for these events is the predilection of the late nineteenth century to pageantry and parade and other forms of public enactment and celebration. As Mary Ryan has noted, this burgeoning civic culture was not devoid of contest and struggle. In her survey of three American cities, she notes that after the Civil War, the forces of order and social control rarely allowed democratic representation to dominate parades and civic performances. Often segregated by race and gender, these public enactments reaffirmed authority, spectacle, and private control. By the 1870s, cities were apparently narrowing the range for the interpretation and expression of the public good. What she calls the "retrenchment of the democratic experiment and the constriction of the public interest" were the long-term results.[16]

If the strategy of civic leaders and politicians was to control through guided tolerance of public expression, particularly during the 1890s with its bitter industrial strife associated with the formation of labor unions, it is also true that a great many urban groups organized parades or participated in organized celebrations on the basis of ethnicity, religion, and national origin. Local residents used such ceremonies to assert their ethnic, neighborhood, occupational, or traditional identities. They often participated in highly visible ways within parades and pageants that ostensibly stressed national unity. This assertion of national membership coincided with the rapid and extensive establishment of patriotic and hereditary societies during the 1890s. By the turn of the century, historical pageantry became even more important, helping to transfigure holidays into celebrations of various American identities and their relationship to the whole. So prominent had such civic celebrations become that by 1913 an American Pageant Association had been organized as a professional organization and clearinghouse for information about the myriad pageants held across the nation.

As David Glassberg has shown in his book on American historical pageantry, historical commemorations in St. Louis during the first decades of the twentieth century revealed an ongoing effort to solidify civic spirit. Especially important, he notes, was the St. Louis Pageant of 1913, marking the 150th anniversary of the founding of the city. It was designed to offset the failure of the city to keep up with its competitor and larger neighbor Chicago and to erase the lingering effects of the bad publicity provoked earlier by Lincoln Steffens's exposé articles on urban corruption in *McClure's Magazine* in 1902. In fact, Glassberg places the Louisiana Purchase Exposition itself within the context of this ongoing attempt of the city fathers to develop and manipulate local spirit through decades of parades, pageants, and other forms of public performance that stressed local solidarity.[17] Of the twelve members of the 1904 Exposition Executive Committee, eleven, including the president, David Francis, were also members of the Veiled Prophet Pageant Society of St. Louis, which held frequent celebrations and parades devoted to exalting the history of the city, promoting the leadership of its local elites, and arousing local patriotism to support reform and civic advancement.[18] These Veiled Prophet parades featured floats with illuminated tableaux fixing St. Louis history into the progress of the world and its most advanced social and technological tendencies. Besides the parades and pantomimes, there was a grand ball featuring the Veiled Prophet himself and hundreds of maids and ladies of honor.[19]

The reform and booster orientation of these commemorations took several forms. Among the participants in pageants, there were often settlement house workers, civic leaders, suffragists, and educators, all of whom believed that a theatrics of democracy would attract the attention and allegiance of large audiences. As Ralph Davol wrote in his contemporary *Handbook of American Pageantry* (1914), "Modern pageantry aims to increase the world's store of happiness by interpreting the meaning of human life and by bringing art and beauty into the minds of all the people."[20]

The American Pageant Movement reached its peak in the years after the Louisiana Purchase Exposition up to the outbreak of World War I. Not only did pageants share with the Fair an emphasis upon public performance and parades, many of them articulated a view of human history that was deeply indebted to anthropological ideas like those displayed at the Fair. Like a great many pageants, performed in small towns and in cities across the United States, they portrayed American history as a mystical evolution from savagery to civilization, from a misty past to the marvels of the enlightened present. Many started with a pantomime or dance characterization of primitive humans—cave dwellers and Native Americans—and then traced the development of America, highlighting the specific city's or town's role in that history. (We see in such performances one solution to the lack of amplified sound and speech.) Anthropological exhibits at world's fairs also presented human history within a similar rubric of evolutionary progress. Although they took place in American cities, pageants often celebrated the ritual entrance of European newcomers to their midst, then the wars that expanded the growing nation, then the building of cities, and then the triumphant creation of an America diverse in its origins but united in one meaning. Or as Davol put it, "The pageant is good propaganda for winning devotion to a beneficent cause."[21] And that cause was most often urban order as a pathway to progress.

Only ten years after the great Exposition of 1904, St. Louis turned again to pageantry to revive the spirit of the Fair and push forward its commitment to progress and civic regeneration (both of which seemed to have faltered). In the masque performed in pantomime before a huge audience (estimated at around a hundred thousand in the local press), citizens reenacted the historical progress of the city, from its Native American beginnings, through French explorers, settlers, and finally, to the coming of the modern city. With the participation of historical organizations such as the Daughters of the American Revolution, the United Daughters of the Confederacy, and the Missouri Historical Society, the St. Louis Pageant Drama Association sponsored these performances and parades for a purpose. Perhaps they were effective in raising public spirit, because shortly thereafter, the city was able to enact a new and controversial city charter. But, like many such celebrations, it denied visible membership in the community to the significant black population of the city.[22]

If pageants and parades were highly ritualized and politically instrumental events during the early years of the twentieth century, stressing national patriotism, local loyalties, racial hierarchies, and ethnic identities in linked concentric circles, their appearance in legion numbers during 1904 on the grounds of the Fair itself should be no surprise. Parade and pageant culture defined the agenda of daily events over the summer of 1904 in St. Louis, and every possible instance was given over to staging a march through the broad avenues between the great open spaces of the display halls or up and down the Pike. Led by blaring brass bands or, often, military units marshaled from the revolving encampments of soldiers on the grounds, these parades featured dignitaries of every sort, as well as representatives from various concessions and displays. Organized to mark the designated days of the Exposition (and every day had its designation or two), whether they commemorated a country, a state, a historical event, or an organization, these parades stirred the fairgrounds into a sea of motion and excitement. Along the Pike, parades often assembled representatives of the native peoples on display, or actors and participants as advertisements for concessions. They both reproduced and altered and extended the contemporary parade and pageant culture of American cities, and at the same time provided a penurious audience, which could not afford entrance into multiple concessions, a chance to view the exhibits in mobile form.

In one specific respect, these parades were reminiscent of the pageants held in the contested spaces of contemporary American cities. Like similar pageants enacting versions of human solidarity all over the nation, public performances on the grounds at Forest Park in St. Louis were carefully controlled by the private management of the Fair. Of course, there were unplanned occurrences, individual assertions, and things gone wrong, but in general the administration was able to direct these events as it wished. With a private police force—the Jefferson Guards—and the power to schedule and control spaces, Francis, with his assistants, exercised a relatively tight control of what appeared in public. Indeed, some observers discovered a model for city administration in the structure of the Fair management. Its efficient and broad planning demonstrated the transformation of politics into bureaucratic management, with control of public order its best example. This form of public Utopia under private management was consistent with the contemporary Progressive Era spirit of reform. Representing a cross-section of the city's civic-minded economic and social elite, the Executive Committee of the Fair was heavily weighted to bankers, millionaires, and members of the St. Louis Country Club, just the group that hoped to impose such a vision of progress.[23]

The selection of St. Louis to host the world's fair, a choice confirmed by congressional approval and a grant of $5,000,000, continued the spirit of holding such events in Western boom-cities. If visitors from Europe (or even New

York) were amazed or amused by the selection, they were simply missing the point of such choices, because all of the Victorian American fairs (save Philadelphia in 1876) were celebrations of this westward movement, claim-staking as it were of an urban civilization spreading out onto the newly settled frontier. As David Francis put it in a speech in 1903, "All the eyes of the civilized world are beginning to turn toward us.... All doubt as to the consummation of our plans, whose magnitude at the inception of the work seemed to tax the credulity of our Eastern brethren, and some of our own citizens, has been dissipated."[24] Among the cohort of western settlements, St. Louis was already an old city, founded by French explorers in 1764 as a fur trade outpost, and whose descendents constituted a sort of self-conscious city aristocracy. Although administered by the Spanish for several decades in the eighteenth century, the city has traditionally honored its longer French ancestry. By 1904 it had a flourishing economy based in manufacturing and transshipment along the Mississippi River and across it with rail lines that laced it to markets in the East and far West. Like many midwestern cities (Milwaukee and Cincinnati, for example), it had a large and prosperous German population, with hundreds of ethnic institutions like Singvereins (choral groups), restaurants, language schools, churches, and German-language newspapers. There were also large numbers of Irish immigrants as well as recent arrivals from surrounding towns and villages and an important group of Anglo-Americans from upstate New York and New England. While Irish immigration was on the rise in this period, the flow of new German immigrants was declining. At the same time, the percentage of African Americans remained about steady or was in slight decline. About 20 percent of the population was foreign born. With about one-third the population of Chicago (575,000 to 1,700,000) at the time, the city fancied itself a serious competitor for commercial predominance and had put in a strong bid for the Columbian Exposition. But its larger neighbor had, in many respects, already won the race for preeminence.[25]

Although the city might have incorporated the Mississippi River, its most important and distinctive geographic (and economic) feature, into the physical plan of the Fair, the administration turned its back on this possibility for a more derivative plan that recalled Buffalo, Chicago, and Paris. It was tempting and far easier to locate the Fair on the relatively level grounds of the far western city land in Forest Park adjacent to the new Washington University and situated in an area that was the focus of recent residential expansion. This choice condemned a grove of ancient trees about which there was some protest. But the Fair leadership promised to restore the area to its prior condition after 1904. As with other world's fairs of this era, there was no intention to create anything permanent: the Fair represented a transitory vision, not a planned expansion of the city, whatever its contribution to growth in the surrounding suburbs.[26]

At the time, Forest Park was much more than trees and a nature preserve. It was also home to Forest Park Highlands, an amusement park opened in 1897. Cousin to more extensive endeavors like Coney Island in New York, the Highlands featured rides, a scenic railway, and vaudeville performances. After 1904, it reerected two new features, a Chinese Pagoda and a tower, both purchased from the Exposition. In the same period, Forest Park also featured sports events and band concerts. In other words, this was already a plausible location for a large fair.[27]

The Exposition was set in a special time in the history of tourism. As a leisure activity, tourism in the late nineteenth century had been descending down the ranks of social and income class. This transition was abetted by special reduced rail rates over the summer of 1904. As the in-house publication the *World's Fair Bulletin* noted in May 1904, a fifteen-day pass from New York to St. Louis was $26.25; from Chicago, $10.00; from Atlanta, $20.85; and so forth.[28] This is both cheap and expensive in 1904: relatively cheap for middle- and upper-class patrons, but certainly out of reach for many working-class potential patrons.

Still, tourism in this era was becoming a possibility for moderate-income families, even to destinations in Europe, and many Americans were taking part in this movement. As Harvey Levenstein has written, there were two fundamental motivations guiding tourism in this era: recreation and culture.[29] From this perspective, world's fairs offered an ideal environment, intentionally incorporating both possibilities and perspectives. There were the elegant, Beaux-Arts exhibit centers featuring an imitative architecture that tourists familiar with Europe would instantly recognize as emblems of high culture. And there were entertainments, restaurants, and performances that appealed to the more recreational side of tourism. The St. Louis Fair purposely incorporated historical elements, reconstructions, authentic objects, and art works in museum-like settings, as well as practical demonstrations of advanced engineering, inventions, and new consumer items. There were copies of structures tourists might otherwise have traveled to see in Europe or Asia (a scale model of the city of Jerusalem, for example in St. Louis). Even the Pike, with its scores of concessions and eateries, mixed the historic and pseudo-educational with venues dedicated overtly to pleasure in a mixture designed to appeal to the desire of the tourist for edification too.

The planners of the Louisiana Purchase Exposition enunciated several broad purposes for this vast undertaking. They focused first on the importance of the fair to St. Louis and the state of Missouri (perhaps because it embodied the personal ambitions of its leaders), then expanded their scope to include the United States and its developing role in the competition among great nations, and finally, attempted to encompass the more abstract purpose of assessing and displaying the progress and potential of industrial civilization itself. This latter, more abstract and conceptual purpose provided the intellectual structure for the layout and appearance of the Exposition, influencing building styles, the positioning of exhibit palaces, and even the arrangement of displays within each building. Both in public pronouncements and private correspondence, the key figures in planning the Fair spoke with an exaggerated vocabulary of self-importance: their purpose was to summarize the best of civilization and to explain the positive dynamics of continued progress for the twentieth century with, of course, St. Louis at the center of it all. As Frederick J. V. Skiff, director of exhibits, put it, "A modern Universal Exposition well might be called an encyclopedia of society, as it contains, in highly specialized array, society's words and works. It is a collection of the wisdom and achievements of the world, brought together for the inspection of the world—for examination and study by its experts. It constitutes a compact, classified, indexed compendium."[30]

But how to organize such a vast encyclopedia? The classification system developed by the fair planners had two distinct purposes. The first was to provide a single universal and orderly grid onto which private exhibitors could place their wares or demonstrate their ideas and institutional innovations. The second purported to demonstrate the relationships among the industrial and intellectual products and processes on display. As the chief of the Manufactures Department put it, the greatest problem was the immense number of objects and the complexity of any mapping scheme that would arrange them in a manageable system. In his final report to the Louisiana Purchase Exposition Company, Director Skiff boasted of a triumph of organization. There were 16 departments, 144 groups, and 807 classes "representing all phases and features of the World's activity."[31] This hierarchical and developmental order reinforced the fundamental evolutionary ideas prominent everywhere at the Fair. Adopting a kind of biological taxonomy similar to the genera, families, and species divisions of the biological sciences, the categories grouped like objects with like objects. Rather than establishing an ecological environment that might show interactions among items, displays of steam engines stood alongside other steam engines, lace work with lace work, and dynamos with dynamos. This categorization was modified in places, as in agriculture, where states could display their products together, but even here, apples and oranges were separated. This organization was so complex in the larger exhibit halls that maps and coordinates were posted for visitors on placards at the entrances. So, for example, in the Palace of Manufactures (Department D), the location of the display of metal corner paper boxes from the American Metal Edge Box Co. of Chicago was: Group 28, Department 363, Block 12-A, at the corner of Fifth and A aisles.[32]

This schema used contemporary theories of biological, social, economic, and cultural evolution to organize industrial and consumer products (even as exhibitors borrowed design tips from the aesthetic logic of contemporary department store display). In the eyes of the planners, the term *evolution* was a key word, and displays were intended to depict development, not just the order of things. As President Francis wrote for *Century Magazine* in June 1904, two prevailing notions determined the display order at the Fair: the possibility of indefinite expansion and development ("evolution if you please," he said); and the importance of process, the "evolution of the raw material to the finished product, the subservience of natural forces to the use of mankind."[33] Given this scientific nomenclature, based upon Enlightenment notions of natural order energized by a Darwinian thrust, it was possible and perhaps even necessary to institute certain forms of privilege of place. Thus, pride of position in exhibit space implied the greatest achievement in evolutionary terms. Not surprisingly, in exhibit halls that had competing national displays, the United States was usually accorded prominence.[34]

This evolutionary principle was likewise the explicit order of the Congress of Arts and Sciences, the St. Louis version of the international intellectual conferences that had become a fixture of world's fairs. Structured according to the guidelines devised by Harvard professor, psychologist and philosopher, and German immigrant Hugo Munsterberg, the congresses were held on the campus of Washington University adjoining the fairgrounds in the late summer of 1904. Like most other fairs of the era, St. Louis hosted a conclave of leading world thinkers, an Olympiad of intellectuals, chosen in this case to demonstrate Munsterberg's professed belief in the unity of all modern knowledge in the arts and sciences. As Howard J. Rogers described it in his 1906 edition of papers presented to the congress, the works were the "literary embodiment of the products of brain and hand displayed elsewhere." In other words, the great encyclopedia of knowledge on display in product and process had its counterpart in the papers given to the conference.

The planning committee, guided by Munsterberg's (controversial) scheme, forced categories of knowledge into his particular vision of modern science, so that Mental Sciences, Utilitarian Sciences, Cultural Sciences, Physical Sciences, and so forth, became the main divisions of knowledge. David Francis underlined the importance of this utility of scientific partition in a speech at the opening ceremonies. As he put it, the purpose was the "unification of knowledge." Even if the notion of unity was controversial and often hard to find in the contentious and revolutionary intellectual world of 1904, aspiration assumed the status of fact in the imaginary university of letters constructed in St. Louis. As Munsterberg proclaimed, the congresses shone a guiding light upon the coming spectacle of unity. The world, he said, "waits for knowledge to fulfill its higher mission."[35]

In fact, Munsterberg encountered a wide variety of problems in putting together his world congress of intellectuals, not the least of which was opposition to his organizational scheme by American colleagues such as University of Chicago sociologist Albion Small. Also problematic was the demand of the French that representatives from their nation have an equal number of representatives as Germany—a reminder of the international competition that broke out elsewhere on the grounds and a violation of the spirit of unity. When the proceedings were published, it was apparent from the papers that disunity prevailed as much as the unity Munsterberg had tried to achieve. Even more discouraging news came later; only around three thousand copies of the proceedings were released, and they apparently did not sell well.[36]

In fact, intellectual unity in St. Louis could only be achieved the way social harmony and order were achieved: through excluding the contentious portions of the world. In this case, Munsterberg did not invite the world of dissonance, nor did the assembled intellectuals represent the full modernist assault on tradition gathering force in the early twentieth century. In the collection of fine arts, the most daring and experimental and important painters were largely absent. So too, was it in the intellectual congresses, for if the brilliant mathematician and physicist Henri Poincaré attended, Einstein, who would revolutionize physics in just a few short years with his papers on relativity, did not; if Max Weber was present (speaking at the "Rural Community" session), no notice was taken of Lenin, who had already published "What Is to Be Done?" in 1902. Nor was W. E. B. Du Bois invited, even though he, of all intellectuals at the time, understood the looming centrality of race questions in the twentieth century, which were, ironically, the major subject of many displays elsewhere on the fairgrounds.

As elsewhere among contemporary world's fairs exhibitions, it was possible to glimpse the unique role designated for the United States to play in the hopeful schema of coming unity. As historian Frederick Jackson Turner (back for a repeat performance after his famous address defining American exceptionalism to the Chicago World's Fair congress in 1893) put it, while all history mattered, of course, the United States had an exemplary role to play, for it was "particularly rich in problems arising from the study of the evolution of society." There were similar examples of such privileged positions ascribed to nations and cultures. Thus, Duncan Macdonald stated in his discourse on Mohammedanism: "The religious is the highest nature of man and among religions Christianity is conceded the place of preeminence, even from the point of view of the purely scientific student of religion."[37]

This universal hierarchical thinking was no better or more obviously illustrated than in the vast Philippine Reservation and the anthropological displays scattered throughout the Fair. Rather than fitting abstract knowledge into evolutionary categories, however, such sections proposed to contain and categorize the entire world of human diversity into an evolutionary scheme in which, once again, the United States occupied the privileged position.

The Philippine Reservation was a large, centrally located compound devoted to displaying the recently conquered and (mostly) subdued island possession of the United States, purchased from Spain as a settlement to the Spanish-American War of 1898 and then conquered through the suppression of an indigenous independence movement. The most prominent element was what must be described as an anthropological zoo, constructed of representative villages from five principal groups brought over from the archipelago. These villages and the native people selected to inhabit them were carefully scrutinized by anthropologists who regulated everything from diet to dancing to maintain what they defined as authenticity. As the *Official Catalogue* of the Fair put it, this was "terra incognito" made legible by the lessons of progress and evolution that could be observed there at work. These exhibits of people and cultures merely extended the general organization scheme of the entire Fair, illustrating in detail the consequence of applying complex hierarchical categories to human civilizations as well as industrial products. The Philippine Reservation was thus a fair contained inside a fair.[38]

As elsewhere in anthropological exhibits, the order and hierarchy of races was first assumed and then asserted. The *Official Catalogue* described the bottom rung of this evolutionary ladder in terms that consigned the most "primitive" peoples to permanent subordination or extinction. As Albert Ernest Jenks, chief of the Anthropology Division of the Philippine Reservation, put it, "Among these collections are those of the small, crisp, wooly haired man, the most primitive inhabitant of the [Philippine] Archipelago, the Negrito; he is a true savage in culture; wandering through the dense mountain forests in search of daily subsistence."[39]

The Fair administration intended the lessons of the Philippine Reservation to be visible elsewhere on the grounds: in the Native American Reservation and in the historical displays in the Anthropology Building. This complex racial discourse included the explanation of the American mission to speed cultural and social evolution where possible in the former Spanish colony of the Philippines. It incorporated a message of nostalgia for the dying, defeated, but now heroic Native Americans who once roamed the Great Plains. It even embraced human identification and categorization in exhibits of the Bertillon system of criminal identification and the new science of fingerprinting. Describing the reconstructed Indian school on the grounds, the Department of Anthropology *Official Catalogue* noted that this section "illustrates the means whereby America bears her share of the White Man's Burden, and the means by which this duty may be done in our insular possessions and in lands beyond the sea." A few paragraphs later, the catalog linked psychometric tests and other anthropological measurements to this overall project of determining "the elements of intellectual character and progress."[40]

It would be well to remember, however, that neither the United States nor the Fair management in St. Louis in any way invented this hierarchical racial discourse, nor did it fit uniquely or even easily into American visions of empire. Indeed, as Diane Coombes has noted in her consideration of Victorian and Edwardian museums and exhibits, the English, especially among others, were actively inventing and constructing concepts of racial hierarchy to justify their vast imperial enterprise. There were, she notes, frequent African exhibits in England that combined science, geography, and anthropology, usually with an emphasis upon authenticity. At the same time, there developed the custom around these exhibits to wall them off, especially "where the proximity of white women and subject peoples threatened to extend 'the colonial encounter' beyond the safe confines of the perimeter fence of the 'native village.'"[41] The St. Louis exhibits participated in this developing culture of primitive display. Perhaps inevitably, there were several controversies that tested the boundaries between imported colonial peoples and the audience that visited the Fair that summer.

The spectacle of humanity on show was just one portion of a larger culture of display that was becoming the hallmark of modern mass culture. Paris, which hosted five world's fairs between 1855 and 1900, was a much-copied site for imagining the modern exhibit world realized in St. Louis. As Vanessa Schwartz has written, these "spectacular realities" defined the meaning of early mass culture in France. American versions of this culture were expressed in newspapers, museums, department stores, parks, and theaters: the living tableaux through which the turn-of-the-century public moved. And just as Paris had innovated in the display of colonial peoples, the vision of Paris and French Beaux-Arts architecture lived on in the United States through the City Beautiful movement that emerged from the Chicago and St. Louis fairs. Something of the spirit of Haussmann's earlier reconstruction of Paris was footnoted in the architecture and layout of both fairs.[42]

The display of human diversity did not end with the living tableaux of anthropological specimens. The Pike, with its myriad amusement and money-making venues, contained a host of concessions and displays that offered anthropology and ethnicity as entertainment, from exotic, Middle Eastern dancing to the popular eateries in the Tyrolean Alps* featuring ethnic foods. In other words, a principal strategy of the management was to present, for purposes of education, entertainment, and profit, the vast array of human civilizations visible everywhere on the grounds and in every format. The *World's Fair Bulletin* (published by the Division of Press and Publicity) put it this way: The Fair anthropology would show evolution from the "Dark Ages of tooth and claw and stone tools, and culminating in the modern enlightenment illustrated in the Great Exhibit Palaces and the International congresses." But this was only the serious version of the prevailing sense that the Fair was also a "wonderful Parade of Peoples," of exotics from the much-photographed "Dwarfs" of the Philippine exhibit to the romanticized depictions of the deposed and famous Indian chief Geronimo, who sold his photograph and other mementos for a fee.[43]

The catch in this monumental effort to organize everything into a single scheme was obvious to F. W. Putnam, the Harvard anthropologist who had tried with only moderate success to organize Chicago's 1893 exposition along similar lines. At stake, he wrote at the time, was the integrity of ethnology, that is, keeping what he called "humbuggery" at bay. The October 1901 *World's Fair Bulletin* reproduced a fascinating letter of advice from the anthropologist to the St. Louis administration. "If carried out in the proper scientific spirit, with a strict adherence to truthful presentation, and no humbuggery allowed from the start, even in the slightest modification of exact conditions," Putnam wrote, "such an exhibit would not only give a distinctive character to the exposition, but would also be one that could never be repeated, owing to the rapid changes that are taking place among all peoples, and particularly among those on the American continent."[44] William J. McGee, chief of the Anthropology Department at St. Louis and former head of the Bureau of American Ethnology at the Smithsonian Institution, wrote that he would endeavor to illustrate human progress from the darkest ignorance to the "highest enlightenment, from savagery to civic organization, from egotism to altruism" using "living peoples in their accustomed avocation as our great object lessons." In his summary report after the event, he reiterated that his purpose was to "show our half of the world how the other half lives," demonstrating that

---

*This concession was variously called the Tyrolean Alps, the Bavarian Alps, and the German Village.

"there *is* a course of progress running from lower to higher humanity, and that all the physical and cultural types of Man mark stages in that course."[45]

As John Troutman and Nancy Parezo have written, the Fair "became a stage on which the anthropologists demonstrated their usefulness to society by ordering the native peoples of the world into a scientifically validated ideological framework that reflected and helped substantiate Western imperialism and notions of superiority in a highly colonialized and racialized world." In addition to repeating a widely circulated contemporary idea, this effort was also an attempt to enhance the position of anthropology as an academic and practical science in competition with educators, religious leaders, missionaries, the military, and entertainment entrepreneurs, all of whom had a stake in characterizing the world's (and the nation's) peoples.[46]

McGee's position in the rapidly developing world of anthropology exemplified the foundation days of the profession. He was a trained geologist turned anthropologist, although his actual fieldwork was limited. Nonetheless, he became prominent in the profession and served as president of the American Anthropological Association and president of the National Geographic Society. His ideas about human racial divisions were a variant on the mainstream notions of distinct differences—a proposition that would remain important in popular anthropology but was increasingly rejected by professionals, although not until after the Fair.[47]

Visitors themselves could perhaps conceive of placing themselves somewhere within McGee's vast tableau of human evolution—that was the intention. As David Francis expressed it: "It will aim definitely at an exhibition of man as well as the work of man.... It will comprehend man in his full twentieth century development exhibiting not alone his material, but his social advance."[48] And just as the problem of humbuggery concerned Putnam in his desire to control anthropology on the Chicago fairgrounds, so too, in a larger sense, was this a concern of the entire St. Louis management as it sought to order and confine the concessionary extravaganzas that encroached upon the more elevated and refined messages of the principal exhibits.

There is something more than a symbolic importance in this struggle of anthropologists to control the exhibit of the "other" at the Fair. They could not, nor did they even try to, suppress some of the most popular entertainments on the Pike, which provided different versions of ethnicity and foreign cultures that only vaguely purported to be authentic. Camel rides, dancing girls, bazaars and markets, foreign foods, and circuses all transformed ethnic difference into commercial intercourse. This popularization of anthropological difference had long been an essential feature of expositions and world's fairs, and they thrived commercially because of it. If, in Chicago, a mistake had been made to mix serious anthropology with commercialism on the Midway, the St. Louis planners sought to segregate the two versions of presentation in two different geographic sections on the grounds. But that could not prevent visitors from confusing the two.

In another sense, the efforts of anthropologists fit the larger and serious project of controlling contemporary American perspectives on race. In the service of imperialism, they sought to reform racial attitudes, to steer them away from two different but equally dangerous sorts of conduct: violence and excessive intimacy. The lesson for St. Louis, for American race relations in general, and for colonial empires was the same: control, enlighten, and contain the interaction between races. Justify the dominance of one ethnic group over another in purely scientific terms. That, it appears, was the deeper purpose of the intellectual commitment to an evolutionary description of civilizations. When we come to closely examine the Philippine display, however, the difficulties and implications of maintaining such a position will become much clearer.

In the world of exhibitions, the exotic and the typical are often closely related. Thus, the Exposition tried to illustrate both. In presenting something idealized, stereotypical, and essentialized, displays often emphasized a schema that assumed the miniaturized and representative could stand for the real—as if the great diversity of the world could be reduced to an encyclopedic compendium of archetypal exhibits. At every turn there were displays of the typical. There were countless scale models, from ideal public school classrooms of the United States, France, and Germany, to a model hardware store, an express office, playgrounds, schools, laundries, libraries, streets, a model Hungarian home, a model college, and model factories. Beyond the model lay the quintessential. The administration seriously undertook to establish this designation by awarding thousands of juried awards for the best exhibits or objects on display from paintings to steam engines.

The ideal of civilization as developing order was also physically reinforced by the way the planners used geography and juxtaposition to assert their contention. As at prior international exhibitions, they gave prominence to the intellectual, cultural, artistic, and technical achievements of Western civilization, placing the vast, ivory sugarloaf display halls devoted to the arts, manufacturing, and the liberal arts around a central core, with halls devoted to machinery and agriculture around their periphery. Next followed the foreign exhibits, state pavilions, the living anthropology of the Philippines, and the mile-long Pike with its shows, parades, barkers, and restaurants. Echoing these endless permutations of intellectual and physical order was the visible presence of social order in the guise of the Jefferson Guards, the staccato blare of military marching bands from several countries, including an impressive group of Philippine Scouts, and the large presence of cadet encampments and troop bivouacs from military schools and militias who were guests of the Fair over the summer.

An aesthetic of contrast underscored these lessons about order. It determined the placement of concessions and curiosities on the Pike and elsewhere on the fairgrounds. An article in the *Atlanta Constitution* of May 1904 pointed this out in a headline about what to expect at the Fair: "Very Queer Things are to be Seen on Every side: Curious Odds and Ends." The largest horse, the most expensive dress, and the smallest humans were all on display side by side. Great distances could be compressed by imagination made tangible: "Every geography student knows that it is a pretty long journey from the Philippines to Jerusalem. For the convenience of the World's Fair visitor they have been placed almost side by side at St. Louis."[49]

The Fair's passion for the classification of knowledge reflected a mood observable in the systematization of ideas and literature under way elsewhere in American culture at this time. In its display at the Fair, for example, the Library of Congress noted its recent achievements in this field of knowledge classification. In 1898 it had developed a revolutionary new cataloging system. Two years later, it substituted movable cards for the ten-year compendium volumes it had previously published listing its holdings. It also established a manuscripts division in 1897. One of this division's first collections was devoted to a history of Spanish possessions: "Special attention is being given to increasing this collection in order that the history of the new dependencies of the United States can be written."[50] For one of the world's great libraries, this drive to organize and classify knowledge was precisely what the Fair was attempting to achieve in its physical collection and display of objects, processes, and ideas. How appropriate was its emphasis on collecting artifacts from former Spanish possessions.

In addition to its larger organizational purpose, the Fair management promoted two other interests: American patriotism and the city of St. Louis itself. In certain respects all of the American international expositions of this era were the work of city-states within the nation, competing with each other, as Chicago and St. Louis did, for preeminence as the trade and manufacturing center for the developing American West. In fact, the previous fairs in this era were all hosted by competing western-oriented cities: Chicago, Omaha, and Buffalo. Much more than a nod to local pride, city leaders in St. Louis understood the enormous benefits that hosting a world's fair could bring. Not only were there immediate economic returns for hotels, restaurants, railroads, guidebook publishers, and a host of small businesses catering to travelers and tourists, but there could be long-range advantages. State exhibits at the Fair also recognized this obvious possibility; they were present in St. Louis to recruit business, immigrants, and tourists.

Depicting St. Louis as a thriving economic and cultural center was as much a part of the planning and execution of the Fair as any other element. Seen in a larger perspective, pride in local place merged into a national vision, and in almost every category, planners made sure to assert that the parochial figured prominently within a larger American patriotism. Local pride, enveloped in nationalism, was the common aim and the reflexive metaphor sprinkled in the speeches of political figures such as David Francis, Theodore Roosevelt, and William Howard Taft and performed in John Philip Sousa's stirring patriotic marches. Even the sculpture attached to the United States Government Building illustrated "the progress of the nation" with its fantasy compositions celebrating "Liberty Triumphant," "Enlightenment," "Liberty Victorious," and "You and the Advancement of the Nation."[51]

Even (or perhaps especially) the poetic doggerel delivered at the Fair delivered the message that St. Louis could dream of a future as the most progressive city in the world's most progressive nation. The "Official Hymn," by Edmund Clarence Stedman, expounded the message in this way:

> O Thou, whose glorious orbs on high
> Engird the earth with splendor round,
> From out Thy secret place draw nigh
> The courts and temples of this ground,
> > Eternal Light,
> > Fill with Might
> These domes that in Thy purpose grew,
> And lift a nation's heart anew.

Or, as this later echo of the national anthem declaimed:

> Oh, say, St. Louis
> > We're looking at you!
> > And so is the whole wide world,
> > As your flag is unfurled
> > Today
> > And, say
> > What do we see?
> > The Future Great?
> > Aw, that's too late
> > You're the Present Great.[52]

This enterprise of patriotism was also a venture for profit. The Fair administration worked energetically to make money through its licensing fees and percentage take of concession profits. It struggled to maintain decorum and counter the impression that St. Louis was either unprepared for the event or that the summer weather was steamy and unpleasant. Because it contained so many different sorts of attractions, the Exposition resembled a vast composite of country fairs, circuses, sideshows, Chautauqua lectures, amusement parks, department stores, and museums, as well as the sleazy inside seam evident in every large American city. It combined and expressed a composite of all these contemporary forms of public entertainment and enlightenment. The large Department of Concessions, with over four hundred employees, authorized the various "bazaars," as it called them, with a double focus on maintaining a balance between propriety and profits.

The planners knew that there was much to be learned from prior fairs, especially from Chicago's Midway experience. They recognized that there was a thriving international culture of world's fair concessionaires. They also realized that they would be adopting displays wholesale from prior exhibitions and amusement parks, such as the Ferris wheel, various Middle Eastern exhibits, the Hagenbeck Circus, the Palais du Costume, the Cliff Dwellers, and the Galveston Flood, among others. The popular Street in Cairo concession, for example, appeared in Chicago and then in the Cotton States and International Exposition in Atlanta in 1895. In a somewhat feeble justification for such duplication, the Department of Concessions "Report" noted how pleasure and instruction might be combined. The visitor could expect to be amused and "more or less" instructed by these displays along the Pike, so long as they did not "degrade." But the real purpose, of course, was to make money for the Exposition.

Norris Gregg, director of the Division of Concessions, reported that Chicago's Paul Blackmer had forcefully justified the inclusion of controversial displays on the Chicago Midway. In a fascinating passage, Gregg included a verbatim record of the conversation right down to Blackmer's most acerbic remarks. When Gregg raised the question of the morality of "Hoochie-coochie" dancing, Blackmer warned him not to censor what some might consider "a National Dance." And when Director Gregg retorted that he insisted upon "a high order" of attraction, Blackmer sarcastically shot back: "Why don't you get every denomination of churches and assemble them on the Midway, that will be a high order."[53]

In their design of the project, the administration and planners of the St. Louis Fair imagined the event within a wide variety of contemporary cultural contexts, from burgeoning tourism to entertainments, to local and national commemorations, to anticipated economic benefits, to imposing some sort of intellectual and physical order on what was a massive bazaar of objects and ideas. So too must historians consider this event within its multiple historical contexts, to estimate how each helped to shape both the nature of the fair and the frameworks within which visitors saw, experienced, and understood it. In doing so, we will inevitably diminish many of its claims to originality, innovation, and grandeur. But we will be closer to recognizing that the Fair was one of many places where Americans might encounter the vast changes that were shaking up modern society and individuals during this era. The significance of St. Louis is that the Fair compressed these possibilities into one space.

CHAPTER TWO

MAKING HISTORY

The management of the St. Louis Exposition went to considerable lengths to shape public perceptions of the Fair, to attract attention and visitors through publicity, and to leave a lasting legacy of meaning for future recollection through its meticulous record-keeping and archival preservation. This chapter explores these efforts and their effects beginning with the publicity department, the story it attempted to set in motion, and its success in placing favorable articles in the national press. Then it moves to evaluate several guidebooks widely sold during the period of the Exposition in which this same message was reproduced and sometimes extended. It then considers two popular novels and their depiction of the Fair. Finally, I will consider accounts written about the Fair by historians whose works often mirror these sources.

While there is now considerable agreement among historians and even an emergent standard narrative about the Fair, this was not always the case. But this fair, as well as other contemporary expositions, have become subjects of considerable contemporary interest and commentary. This chapter will also explore ways in which the views of Francis and his fellow administrators have found their way into historical accounts, primarily through the rich legacy of company records and other archival sources. Although historians today are profoundly critical of the intentions of the Fair builders, particularly for their exploitation of subjected peoples, they nonetheless continue to reproduce their words and ideas as explanations of the meaning of the event and to privilege this meaning. Thus, David Francis and his colleagues continue to define the significance of the Fair by virtue of the legacy of their administrative records.

David Francis not only gave innumerable speeches on the fairgrounds at every special opening and commemoration, he also spoke widely in St. Louis and elsewhere in the United States about the Fair, lending his considerable national visibility to the effort. His ideas became the basis of the immense campaign of the publicity department. Added to this were his history, *Universal Exposition of 1904*, published in 1913, and then his activities as the first president of the Louisiana Purchase Historical Association, devoted to preserving the memory and records of the Exposition. Inevitably, these efforts are responsible for much of what we think about the Fair today.

In an article published in *Century Magazine* early in the summer of 1904, Francis extolled the purposes of the Fair from its planning stages onward: "first, that provision be made for the indefinite expansion and development—evolution if you please—of the exhibition; second, that so far as practicable the Exposition be an exhibition of processes." With this latter phrase, he meant that exhibits would be encouraged to demonstrate manufacture, "the subservience of natural forces to the use of mankind"[1]

In the *Universal Exposition of 1904*, Francis wrote of how this purpose was grandly reflected in the accomplishments of the Fair. It was, he declared, "the scene of the exhibit of the best effort of man from the beginning of society." This broad definition of evolution, which placed St. Louis at the pinnacle of progress and the United States as the host of the best hopes of mankind, declared the unity of civilization around notions of progress and industrialization. As Francis wrote (with the bias of his age), the Louisiana Purchase Exposition was also noteworthy because of its "distinct universality and democracy. No discrimination between race* and sex was shown. All were entered on an equal footing. The work of woman was exhibited alongside that of man with equal prominence; the manufactures, industrial and fine arts products and educational methods of the Americas with their composite races, of the various nations of Europe with their varying bloods and of Japan and China were presented in competitive alignment as the products of citizens of the world."[2] Francis also emphasized the commemorative circumstances of the Fair, the anniversary of the Louisiana Purchase, which opened up the dramatic physical expansion of the United States. Beyond this larger purpose, of course, was the hoped-for boost to the St. Louis economy by demonstrating the administrative skills and convenient facilities of the city for receiving visitors and commercial investment. An unspoken mission

---

*A discussion of racial exclusion follows in chapter 5.

was to counter Lincoln Steffens's disheartening revelations of corruption and political malfeasance in 1902.[3]

Aside from celebrating all that was innovative or original in the exhibits (light bulbs, refrigerators, phonographs, motion pictures, diesel engines, radio, telephones, etc), Francis also expended considerable energy talking about the ethnicity and anthropology exhibits at the Fair. It is worth exploring his actual prose on this subject:

> The Philippine exhibit furnished a complete exposition of the slow evolution of civilization among tribes comparatively adjacent to but isolated from each other when left to their own resources, though in the region of the oldest civilization; also of the effect of contact and mixture of certain races with Caucasian blood and western civilization. A group of African pygmies, some of them of man-eating tribes, brought the collection of primitive races down to the lowest know human stage; the Ainus ["white" inhabitants of Japan] and Patagonians represented the semi-civilized savages still existing like our Indians in countries dominated by highly civilized races and like the Indians rapidly disappearing.[4]

Several elements in this paragraph merit explication. Most important is the centrality of the Philippine Reservation to the planning and execution of the Fair. Francis is clearly repeating received anthropological wisdom about the evolution and categorization of races and ethnic groups, even down to the notion of disappearing races. It is also evident that he understood the recent acquisition of colonial dependencies in the context of America's continental expansion into lands made available by the Louisiana Purchase.

The source of his ideas about anthropology was, no doubt, William. J. McGee, of the Bureau of American Ethnology at the Smithsonian Institution. As head of the important Department of Anthropology, McGee wrote and spoke widely about his purposes of exhibiting people in their "natural" setting. "The aim of the Department of Anthropology at the World's Fair," he noted, "will be to represent human progress from savagery to civic organization, from egotism to altruism [using] living peoples in their accustomed avocations as our great object lessons."[5]

Instant "histories" of the Fair stressed many of these same ideas but not always with the same purported scientific detachment. For example, the extensive *Book of the Fair* by Marshall Everett, published in 1904, featured a hodgepodge of photographs set in an eclectic order. Indeed, Everett's book failed to reproduce the careful and controlled anthropological vision that McGee and Francis insisted upon in their explanations of anthropology at the Fair. The author was also careless about expressing his prejudices and frequently deployed ethnic slurs. Everett juxtaposed varieties of native peoples to make his points, sometimes with humorous, if insulting, intent. Describing a visit of Chief Yellow Hair of the Sioux and his warriors to the Philippine enclave, he noted that Native Americans were horrified to discover that men worked while women sat in the shade and gossiped.

> Looks of disgust came over the faces of the visiting braves. All said, "ugh" with unmistakable displeasure.
> "What matter?" was asked by somebody who had an idea that an Indian could understand better if addressed in chinks [*sic*].
> "Men work, ugh!" answered Chief Yellow Hair, and he and his braves strode majestically out of the Bontoc Village and returned to their camp.[6]

Everett's rich vocabulary of bias became a glossolalia of prejudices in his account of the Philippine Igorots.† For example, he noted that a local schoolteacher tried to teach young Igorot children to read. She failed dismally at this task until she struck upon a word they readily learned, *dog*. This story was a scurrilous reference to the reputed diet of the ethnic group. He also highlighted the "vicious and bloodthirsty" habits of the Moro group, referring to their supposed cannibalism. As he moved up the scale of "civilization," he began to detect individual differences among Filipinos. In general, though, his conclusions were discouraging and disparaging: "the most striking fact brought out about the culture of the wild peoples of the islands is its shallowness." Worse, their dismal state seemed to argue against the American civilizing mission, a prominent ideological element of the Philippine display.

Prejudices vied with admiration in his long section about the Japanese. As he noted in one conflicted passage: "Yes, the Japs are a wonderful people, and at peace or at war, at the world's fair or at home, they have been written about over and over again. But, somehow, always there is something more to be said. If it is not a vase, it is a fan; if not a fan, it is the silk; and if neither, it is the quaint ways and sharp cleverness of the nervous little fellows themselves."[7] This is not the sort of discourse that the Fair encouraged. On the contrary, Everett's words suggest the sort of common prejudices against which the anthropologists and their publicists struggled.

Contrast this with the statement of Alfred C. Newell, chief of the Division of Exploitation (overseeing concessions and other forms of commerce):

> About the time the World's Fair City is waking at early morning, one hundred bare-limbed Igorots often sacrifice and eat a dog on the Philippine reservation. At the same hour, scarcely two hundred yards away, a bugle sounds reveille, and

---

†The spelling of this Filipino group varies widely in the sources. I will use *Igorot* except in quotes.

four hundred well-trained soldiers in the blue of the United States Army hustle from their tents. These are the Philippine Scouts.... All of these people live on the same island in the Philippines. The Igorot represent the wildest race of savages, the scouts stand for the results of American rule—extremes of the social order in the islands.[8]

Attempting to control and shape reactions to the Fair was a monumental task, but the administration expended Herculean efforts to that end. It published the daily broadsheet entitled *World's Fair Bulletin* to call attention to special performances, displays, and meetings. It also released *The Piker*, a periodic journal designed to highlight events, exhibits of special interest, and concessions along the Pike. The journal was aware of the inevitable competition and comparisons of paying attractions with the more sedate exhibits of the grand display palaces. So, in May, for example, it noted that the "Pike seeks to elevate the dominion of entertainment, [but] it loses none of the spice or that exhilarating feeling which the public has come to expect in this feature of an Exposition."[9] The next month's issue compared the Pike with Coney Island in New York, claiming that the Pike provided a more current and respectable entertainment center.

The Fair administration undertook an enormous effort to publicize the Fair and fix, in the minds of potential visitors, a compelling impression of the Exposition. The task of the Division of Press and Publicity was also to discourage disparaging rumors, for example, about St. Louis's notorious undrinkable water and lack of acceptable accommodations. Employing thirty newspaper men as well as a team of stenographers and other employees, it turned out thousands of press releases. Likewise, the division courted visiting writers and issued twenty thousand free press passes to reporters. The results were impressive. Daily papers in St. Louis alone published thousands of column inches of reading matter about the Exposition. During the first six months of the Fair, over 3,000,000 column inches of articles appeared in English outside of St. Louis, with 160,000 in German and 80,000 in French. Almost seventy general and specialized trade journals devoted special issues to the Fair, and many of them reproduced its press releases verbatim or made them the basis of news stories.[10]

The division also worked diligently to bring national and international congresses and meetings to the fairgrounds. It succeeded in attracting 428 officially recognized groups, including the Democratic National Convention (for nominating their presidential candidate) and the annual meeting of the National Education Association. With advertising circulars placed in hotels across the country and on trains and in stations and even stereopticon slides prepared for roving lecturers, the division garnered wide attention. It also paid

TABLE 1. Articles on St. Louis World's Fair Identified from *Readers' Guide to Periodical Literature*.

	Number of articles		
Category	1900–1904	1905–1909	Total
Machinery and inventions*	29	1	30
Art	14	8	22
Education	6	7	13
Foreign nations	12	0	12
Anthropology			12
Race and anthropology†	8	1	9
Philippines	3	0	3
The Pike	5	0	5
Agriculture	4	0	4
History	1	0	1
Religion	1	0	1
Grounds	1	0	1
Views	1	0	1
Buildings	1	0	1
History	1	0	1
Women	1	0	1
Total			105

NOTE: This table is based on titles of articles compiled by the *Readers' Guide to Periodical Literature* in their 1900–1904 and 1905–1909 cumulative volumes. *Readers' Guide to Periodical Literature*, vol. 1: 1900–1904 (Minneapolis: W. H. Wilson, 1907), 1268–1270; vol. 2: 1905–1909 (1910), 1943.

*Machinery and inventions are listed together because of their close proximity in the display pavilions.

†Race and anthropology are listed as a single subset because, in almost every case, the subject of the articles on anthropology focused on defining racial characteristics.

special attention to the area around St. Louis, within 100 to 150 miles because, as its final report noted, "from sixty-five to eighty per cent of the paid admissions were drawn from within" that radius.[11]

My own survey of articles cited by the *Readers' Guide to Periodical Literature* for the period 1900 to 1910 suggests something about the primary topics of these articles, suggesting, in other words, what the principal mass circulation magazines thought would catch the attention of their readers. Arranging these articles by the topic named in the title reveals that the most important subjects were machinery, art, education, and foreign nations. The combined topic of race, anthropology, and the Philippines tied with foreign nations as the fourth most frequent subject.

Examination of the content in many of these articles indicates that they are, in general, celebrations of the Fair and advertisements for a visit. *Cosmopolitan*'s special issue of September 1904 was perhaps a bit late in urging tourists to travel to St. Louis, but the articles cover a comprehensive range of subjects and reveal the power of suggestion exercised by the publicity department. The editor, John Brisben Walker, spent eleven days as a guest of the Exposition. The result was several articles on almost every aspect of the Fair, except for one important omission: the Philippine Reservation. Although Walker suggested that visitors devote a half day to studying anthropology in the building housing special exhibits illustrating that science, he did not discuss the Filipinos or other displays of living peoples. Fascinated by technological anthropology—objects, tools, weapons, and so forth, he did not dwell on the living exhibits that McGee had gathered together. This focus on material culture coincided with his larger concentration on different categories of the exhibit that emphasized science, invention, education, the progress of women, and new consumer products—in other words, the economic and cultural developments of contemporary industrial society. He also stressed elevating entertainments such as The Story of Creation and the scale model of the walled city of Jerusalem near the Pike. Indeed, perhaps half of his articles dealt with displays and concessions on the Pike (not, of course, the "salacious entertainments") as well as historic reenactments such as the Boer War and the Naval Battle of Santiago (from the Spanish-American War).[12]

The *World's Work* was another national journal that devoted a special issue to the Fair. Like *Cosmopolitan*'s, its coverage was idiosyncratic. The most important advice given by the editors was to see the Fair in a "logical" order. This logic more or less conformed to the Fair administration's emphasis upon the theme of process. Following from this, the magazine constructed an orderly presentation that began with power generation, transportation, agriculture, and mining and ended in manufactures and fine arts. While the journal suggested examining the Philippine Reservation, it did not dwell on the overriding questions of race and evolution that structured the arrangement of that exhibit. Finally, the journal advised a visit to the national exhibits of Germany and Japan.[13]

Other national journals tended to present the Fair in ways that would appeal to their special-interest audiences. Thus, architectural journals stressed the order, style of buildings, and larger issues such as municipal improvement. General circulation magazines tried to appeal more broadly. On occasion these articles were simply endorsements. The *Independent*, in May 1904, published an article by William Saunders, the secretary and general manager of the Business Men's League of St. Louis. He tendered advice about how to see the Fair. He suggested that visitors take two steps to orient themselves: first, enter through the main gate so as to see the whole Fair ensemble from its best vantage point, and then take a quick trip to their state building to find their local bearings. After thus orienting themselves, he advised, visitors should plan to spend two hours in each of the main exhibit palaces, paying special attention to electric presentations. A day at the Philippine exhibit, time attending athletic performances, and then a trip to the Pike rounded out his recommendations.[14]

The *Nation* magazine was far more ambiguous about the Fair, turning it into the pretext for a political discussion. Nor were the editors entirely enthusiastic about St. Louis, although an early article expressed surprise at the immensity and beauty of the grounds. Its author, Mabel Loomis Todd, noted that the "Eastern papers" had carried little in the way of a physical description of the grounds. There was, she noted, much to be seen, although her interests concentrated on the fine arts to the disparagement of the Pike, which she dismissed as an "expurgated Midway."[15]

In a later summary evaluation, one of the *Nation*'s correspondents expressed fair fatigue and the damning (and valid) charge that other expositions had presented many of the same exhibits but with more novelty in the original version. An article marking the closing of the Exposition suggested that St. Louis was suffering serious financial difficulties. But most interesting was the magazine's political assessment of the Philippine exhibit. Allied with the "Mugwumps" (a political designation for Republican reformers, many of whom opposed seizing the former Spanish colony), the *Nation* concluded that its readers might nonetheless find solace for the "Fathership of the Philippines . . . imposed on us" in the unambiguous success of American education in the archipelago demonstrated in the model school at the compound.[16]

In other instances, there was a different mixed message about the Philippines. Walter Williams, writing in *Outlook* in August 1904, embellished a cliché developed about other world's fairs: that they were universal samplers of culture and could stand in for a trip around the world. Consequently, he portrayed the huge variety of civilizations on display at St. Louis, the wide variety of indigenous forms of religious architecture, and the appearance of models of famous buildings as a world of travel condensed into one place. Half tourist and half anthropologist, he indicated that the visitor would find an immense variety of odd and interesting cultures. As for the Philippines, he voiced two common reactions. First, he invoked the stereotype of primitiveness: "The unclad Igorrotes dine on dog and lament the lack of head-hunting expeditions." On the other hand, he praised the huge sum expended to "show the wonderful present in the Philippines and the prophecy of the yet more wonderful future."[17] Would visitors be more intrigued by the grotesque attraction of cannibals and dog-eaters or by the demonstration of strides made through imperial conquest? This was a question that hovered over the whole Fair: Would patrons and readers get the message about serious anthropology? It is not clear from reading this journal literature which was really the case.

Given the opportunity to place his own article in *Century Magazine,* David Francis extolled every exhibit at the Fair, hoping to generate enthusiasm for a visit by offering something for everyone. Overall, he conveyed the theme of order and classification that was the message of many of the press releases written by the publicity department. All of the exhibits, he noted, were subject to "vigorous and impartial selection" that stressed harmony with the larger purposes of the Exposition. All this, he said, was submitted to a classification system "acknowledgedly [*sic*] the best that human intelligence has yet devised." Even the Pike, he suggested, had been "elevated from a Midway aggregation of novelties to a dignified connection with the Exposition itself." When seen as a whole, the Fair was a valid demonstration of process: "machinery in motion, instruments in actual use, illustrating the evolution of the raw material to the finished product." This was the grand purpose of the classification system: to establish and celebrate the imposition of order upon nature, human production, and society and its unruly inhabitants.[18] This grand educational endeavor included not just the model Indian and Philippine schools and other demonstrations, but, in a larger sense, the proposal to educate the public about relationships inherent in the new industrial society emerging at the turn of the century among manufacturing, consumption, design, and urbanism, and on a larger scale the connections among various cultures and societies of the world.

As might be expected, the national newspaper press picked out items of local interest for their readers or focused on the odd and unusual at the fair or, sometimes, the visits of political celebrities and foreign dignitaries to the grounds. Particularly true of St. Louis papers, but also true elsewhere, the stopover by the U.S. president's daughter, Alice Roosevelt, in the city occasioned a noteworthy event. The St. Louis press also published an extensive accounting of events on the fairgrounds, including stories about visitors, openings, and unusual features. It recorded the controversy over segregation in restaurants and other concessions on the Pike and rehearsed stories about disputes among the Igorots and Filipino Scouts,[19] who were, apparently, flirting with Igorot women. There were also articles about Filipino men escorting white women and reports of the fury directed against them for this transgression of racial propriety.

The *St. Louis Globe-Democrat* published extensive accounts about the Philippine exhibit but rarely expressed much interest in the complexities of anthropology. A June article about press humorists invited to the Philippine exhibit suggests the flavor of these pieces. "A special programme had been arranged at the Igorrote Village," it noted, "including a series of spectacular dances and a dog feast. The latter spectacle appeared anything but funny to the joke manufacturers, many of whom took to the woods. The savages had been warned beforehand that their efforts would be subjected to humorous comment. The dog eaters took pains to make the feast as revolting as possible." Accompanying the piece was a photograph of a crowd of visitors ranging very close to a group of Filipino men wearing only loincloths. Lurid and disgusting to an audience prepared to be simultaneously titillated and repulsed, this spectacle was only the logical result of a display that was torn between education and the thrill of the exotic. But the press generally had no difficulty choosing the latter to depict.[20]

The numerous guidebooks published for the occasion of the Exposition served as another source of information. Unlike newspaper or journal articles or occasional pieces and advertisements prepared by the publicity department, however, the guidebooks were intended for special use. Prepared on the basis of information provided by the administration, they faced peculiar time constraints. Most, if not all, contained an extensive photographic program, a sampling of sights, curiosities, and panoramas. To be useful, though, they had to appear early in the tourist season; otherwise they would not be competitive. This resulted in a wide variation in quality. Some guides contained advice and photographs based upon knowledge gained once the Fair opened; others attempted to divine what would be the most interesting exhibits before they were installed. Some featured photographs obviously taken before the official opening, when buildings were still under construction and the grounds barren and without landscaping.

But guidebooks could function in two primary temporal contexts, one before and during, and one after the Fair. Used on the spot, they proposed a variety of itineraries through the fairgrounds, gave advice for travelers, discussed the history of St. Louis, and provided other useful information. They could be used to plan and execute a visit. But they could also be critical afterward in structuring memories of St. Louis and the Fair, lending coherence to what in many cases had been a helter-skelter meandering through the grounds. Guidebooks also became a part of the information available to novelists who constructed imaginary trips to St. Louis in 1904—a literature that had become widely popular around the Chicago Columbian Exposition in 1893 and again in 1904.

One of the (universal) problems of guidebook writing was to describe in extravagant fashion the unusual, the odd, and the remarkable at the Fair and, at the same time, provide the visitor a sense of comfort and safety within this strange environment. Two examples illustrate how this was accomplished. In *St. Louis and Its World's Fair,* German author Hermann Knauer designed an English-language book for German tourists—either from the Continent or for the many German Americans living in St. Louis or elsewhere in the United States. He particularly stressed all of the exhibits with distinct German contributions and spent the largest portion of the guide discussing the Tyrolean Alps, located on the Pike. This was hardly a surprise, since the author was also the designer of the concession.[21]

But Knauer's strategy of focusing upon the anticipated identity of the reader was repeatedly invoked by other guidebook writers who advised an initial orientation according to the familiar: state pavilions for their residents, agricultural exhibits for farmers, and so on. This attention to the possibility of disorientation is a fascinating element of otherwise standard texts. For example, in his somewhat disorganized, privately printed compendium of fair attractions, *Sherlock's World's Fair Guide and Bureau of Information,* Eli Sherlock, gave some odd but interesting pointers about how to maintain equilibrium when faced with the extraordinary. Acknowledging that it was impossible to see most of the exhibits, he wrote: "It is important therefore to first examine that which pertains to you or your own vocation or calling. For instance, if you are a farmer, see the agricultural, horticultural, live stock and kindred subjects." Warning against any "frivolous or cursory examination of exhibits," he continued that one might visit foreign buildings or the Pike, but, he concluded, "Hang on well to the exhibits which represent your own trade or occupation."[22]

More mainstream guides confronted the huge array of possibilities on the fairgrounds with detailed itineraries. For example, the *Rand-McNally Economizer* proposed a number of day-trips based upon the visitor's length of stay. For its six-day tour, it began with the principal large buildings that occupied the center of the Fair, such as those devoted to manufacture, liberal arts, and education. This preference for high culture and educational exhibits extended through the entire guide, ending finally in an eventual descent into pleasure on the Pike put off until the last day. Even within the Pike, however, the guide suggested a visit to the Tyrolean Alps and the Irish Village and then a tour of the model city of Jerusalem. Along the way, on day five, for example, the guidebook suggested a visit to the Philippine Reservation, anthropology exhibit, and physical culture exhibit. This rather typical emphasis upon the educational elements of the fair, consigning the Pike and its amusements to the end, or failing to discuss them at all, is typical of many of the principal guides to the Fair.[23]

Another Chicago company, Laird and Lee, published a *Standard Guide* to the Fair that paralleled the order of the Rand McNally guide. Beginning with a history of the Louisiana Purchase, the guide then advised visitors to see the major exhibit buildings clustered around the center of the exhibition—U.S. Government, Agriculture, Horticulture, Fine Arts, Liberal Arts, and Manufactures—and then to move on to other exhibits such as anthropology. It advised visitors to make lists of things for "all members of family and friends" to do and to set aside days to see them. Implied in this advice was the suggestion to make a long and thorough visit as well as to consider the variety of interests in the group.[24]

A local publication by W. W. Ellis, *The Piker and World's Fair Guide*, repeated the slogan "Processes as well as products," which had been adopted by the administration as the theme of the Fair.[25] Like other guides, this one also suggested a six-day itinerary (not seven, because of Sunday closing). Ellis included a long biography of David Francis as well as considerable information about the intellectual congresses and meetings of national groups, such as the Women's Christian Temperance Union (WCTU) and the National Creamery Men. And like most of the guides, this one also had its eccentricities: Ellis included the local baseball schedule, a fishing guide, and trap shooting rules.[26]

Another St. Louis publication, the *World's Fair Authentic Guide* (endorsed by Francis, it claimed), attempted to structure the experience starting with the first impression, instructing the visitor to notice the sublime quality of an initial view: "Nothing so big, nothing as beautiful, has ever before been attempted by man." This guide suggested a variety of itineraries for two, four, six, and ten days. It repeated the publicity department's description of the Exposition as a "universal cyclopedia of society, being a living picture of the artistic and industrial development at which mankind has arrived." In fact, the guide was an unusually thorough compendium of information about the Fair gathered from facts and figures supplied by the administration. Classification, anthropology, the Pike, all the departments of the Fair, music, statuary, special performances, and national meetings were each covered by this thorough pocket book. If there was a comprehensive guide to the Fair, reflecting the diversity of exhibits and the grandest purposes of the administration, this was it.[27]

The guide advised a special visit to the eastern section of the Manufactures Building. There "is an exhibit that all the ladies and most of the men will be interested in. It is the sumptuous display of gowns, not draped over dummies with wax faces, but on the living model."[28] This interest in fashion, dress, and costume is one of the major elements of the photographic program of guidebooks and the photo souvenir publications published as memory books for a national audience that did not travel to the Fair. Most guidebooks included photographs of buildings and perspectives, paintings and statuary. But their most interesting reproductions were of people.

Although I will discuss this more extensively in a chapter on photography, it is essential to underscore how these photographs added to the depiction of the Fair as a celebration of Western history and progress. Costume and dress were key elements of this presentation. Photographs captured visitors in more or less formal attire: men sporting caps or hats, white shirts, sometimes vests, and long, dark trousers; women in long skirts, shirtwaists, and large hats, sometimes carrying parasols against the heat. Anyone perusing a guidebook or photo publication would see these pictures of the audience, many fashionably dressed, looking at the exhibits and at other people. The objects of their gaze were, in general, decked out in some sort of costume: native dress, ceremonial outfit, or uniform of one sort or another, but in every case, some article of clothing signifying that person's identity as part of a special and (unchanging) culture or special social function. This contrast in dress was a key element in the representation of the Fair to the larger public and a way of inviting the purchaser of the book to become part of an audience looking at others, rather than being observed. The creation of a subjective photographic persona in these guides, identified as an observer, even if it was only visually suggestive must be accounted a crucial element in the establishment of perspective and orientation to the Fair, as important in some respects as the verbal instructions. It could be as significant as advice about how to see a building or what conclusion to draw from an exhibit. In this respect, photography underscored and exaggerated the suggestion that paying visitors and guests of the Fair were attending a grand cultural performance. Furthermore, the sequence of photographs in guide and souvenir books, moving generally from high culture buildings around the core of the grounds first and to the Pike or ethnological exhibits last, subtly affirmed that this was also an order of values. These books visually depicted a movement downward from the most important exhibits to the least, the visualization of evolution glimpsed backward through a devolution of images.[29]

Two fascinating and widely sold novels about visits to the Fair were written primarily on the basis of guidebooks and official releases from the managers of the Fair. Both authors, Charles M. Stevens and Marietta Holley, used the popular format of dialect novels to create humorous encounters by ordinary American folk with the remarkable and unsettling possibilities of the Fair. Both authors wrote multiple world's fair novels, with their main characters visiting Chicago in 1893 and then St. Louis in 1904. (Stevens also used the Panama-Pacific International Exposition of 1915 as a third setting.)

Stevens, a popular history writer, encyclopedist, and editor of the *Official Guide Books for the Louisiana Purchase Exposition,* resurrected his Uncle Jeremiah character from a novel about Chicago's World's Columbian Exposition. Jeremiah led a motley group of country rustics from "Strumpburg" (Indiana) to St. Louis for a week of comic escapades. Among the seven persons in the group were a German immigrant, an Irish immigrant, two children, two teenagers, and, of course, Jeremiah. The plot, such as it exists, is the story of the various encounters of the group with strange modern technology and particularly with the exotic peoples gathered for display. This is enhanced by a romantic plot that eventually ends with the marriage of two characters, Nora and August.

In most respects this novel resembles other humorous accounts of encounters of rural Americans with the burgeoning cities of fin-de-siècle America, and it should be read as part of that literature. The Fair just exaggerated and compounded the possibilities for cultural misunderstanding and pseudo-danger typical of these stories. Inevitably, the group most frequently visits the Pike and, of course, the native reservations of Filipinos and Native Americans. Nakedness (and sexual temptation), confusion, and danger from con men and swindlers are all typical elements of this genre, and Stevens thoroughly exploits them. Visits to the Igorots and other ethnic groups are occasions for jokes about their strange customs. While no ethnic group seems to escape satire (including, of course, the innocents from rural America), in the end, Uncle Jeremiah makes a plea for tolerance. He imagines Americans on display, with a barker pointing out their peculiarities of dress and custom. As Uncle Jeremiah says, the lesson is that "human nature is the same all over the world, and it is the most variegated curiosity on earth."

Uncle Jeremiah is also capable of suddenly stepping outside his dialect persona to expound about important matters in guidebook lingo. So, he notes, "For surging millions of happy people, it [the Fair] is a silent and supreme industrial university in the achievements of mankind. A magnificent spectacle of the triumphant will of human evolution, a mighty epic poem composed by millions through the long course of historic ages."[30]

What might a national audience have learned from such a novel? How might attitudes be shaped by such a story (Stevens claimed sales of 475,000 copies of his Chicago book)? Given the book's format as a genre novel about rural-urban encounters, the Fair was, in fact, only a backdrop for a much more familiar story of the escapades of rubes and city slickers. Stevens did include one fascinating justification of the state pavilions that could be seen as advice to visitors. The buildings, he says, were "club houses for us bodies who get weary so easy, and as a public hall for our politicians to show off." And there were a number of descriptions of the buildings, statuary, and displays on the grounds. But most of all it was the story of cultural misunderstandings on the Pike and in the anthropology exhibits, resolved gently and with good humor.[31]

Marietta Holley's widely read Samantha series also accompanied a dialect speaker on a tour of the St. Louis Exposition. According to her biographer, Holley's books were widely read, and apparently so, because she published a variety of Samantha stories, including travels to the South, to Chicago in 1893, and to St. Louis in 1904. Holley also wrote widely on a variety of subjects, including race issues and women's suffrage. She, herself, however, did not visit the world's fairs to research her books. Instead, she relied upon guidebooks as her source. The result is similar to Stevens's work, with descriptions and information presented as if lifted verbatim from these guidebooks and offered in passages of stilted standard English that jar with the dialect speeches of the main characters.[32]

Also like Stevens, Holley used her main characters to express a sentiment of innocent awe at the spectacular buildings and displays. Much of the book, in fact, is given over to long descriptions of the buildings and their interiors. What is most striking, however, is Holley's ambiguity about race and imperialism. She introduces the character of a black servant, Aunt Tryphena, whom she treats satirically. Yet in an extended and serious section, she muses on the meaning of race and ethnicity in America. During a visit to the Philippine Reservation, Samantha thinks to bring up the Civil War and the plight of freed slaves—an interesting association in itself. She sees a direct relationship between American race relations and colonialism: "I hope Uncle Sam will do well by all the folks he's garden [sic] over," Samantha says, "the Indians, Negroes, Phillipinos [sic] and all, I believe he means well by the hull on 'em, but he has so much on his hands, he don't know which way to turn."[33] On the other hand, she expresses optimism about the possibility of Uncle Sam turning "savages and cannibals that inhabit part of these new possession into good American citizens." For American blacks, the resolution was less clear: perhaps a special agricultural enclave in Florida or emigration back to Africa. And yet Samantha demonstrates genuine sympathy for America's "steal children, the dark complexioned ones stole away from their own land to be slaves."[34]

Reading about the escapades of Jeremiah and Samantha could impart considerable incidental information about the Fair, much of it taken from other printed sources on the Exposition. In general, this information followed the outlines of publicity developed by the Fair's managers. But both novels spent considerable time exploring cultural encounters between an idealized American character (Samantha and Jeremiah) and the culturally exotic peoples they met on the grounds. Both used thin romantic plots to move their stories along, but the real subject was the humor to be generated by misunderstanding and crossed communications. The Fair, in these authors' imagining, was a reproduction of the new urban world with its well-known perils and temptations but with the added confusions of a new international world brought together in a kind of cultural sampler seen along the Pike, in concessions like the Street in Cairo, and in the ethnological reservations and displays. The tone was one of wonder, fascination, and puzzlement. But the puzzlement and problems of these fictional fairgoers were never strong enough to discourage a potential real-life visitor, for these novels also belonged to the genre of guidebooks describing an imaginary visit. Ordinary Americans—very ordinary Americans—they seemed to say, would find amusement in a trip to St. Louis. By and large they followed the narrative of grand exaggeration outlined by the Fair administration. As encounters with different cultures and societies, however, both novels placed that experience within the more familiar story of a trip to the city, the archetypal story of American late nineteenth-century urbanism. And perhaps most important of all, they placed the story of family at the center of their narratives.

Stories of the Fair developed by David Francis or by the publicity department, were reproduced in magazines and newspapers, in the guidebooks, and in the Jeremiah and Samantha novels. Today they serve as essential sources for our understanding of the history of the Exposition. But we need to recognize that this material has followed not just one, but several different narrative schemes, some reproducing genres popular in other formats, some seeking to orient the visitor within familiar boundaries before the visitor ventured out into the Fair itself, some appealing to special interests by occupation, and some merely entertaining. Even the publicity about the Fair was not entirely consistent in what it stressed. Nonetheless, a select few of the purposes expressed principally in official documents and speeches have emerged as the principal historical narrative. Of course, historians have restated the ethical and cultural positions of management in their own, highly critical, moral vocabulary. This means that the racial evolution and hierarchy promoted by the administration has come to dominate the historical meaning of the Fair but in a negative mode. Writing within the terms of contemporary moral discourse, historians have concluded that the Louisiana Purchase Exposition *meant* the promotion of a racialized anthropology and the justification of imperial conquest.

My reasons for talking about historical works in the same chapter as guidebooks, publicity, and novels more than hint that I have serious questions about the way sources have been collected, preserved, and used by historians.

To begin with, there is our emphasis on elite, archival sources which has, for better or worse, allowed the creators of those sources to continue speaking through the historical record. Their purposes continue to stand for the meaning of the event. Theirs is the voice of history. And in assuming a perfect correlation between intention and reception, we have privileged their vision, however much we might personally deplore it.

In writing the history of events such as the St. Louis Exposition, historians implicitly or explicitly have to make a number of decisions about strategy, presentation, and meaning. First is assessing the importance of such an event. Was it central to the historical period? Does it influence or perhaps reveal important opinions or cultural developments? What story does it tell, or rather, which of its many possible narratives is the most significant? What elements should the historian emphasize? And, perhaps most important, what are the contexts within which the event should be understood? There are two further questions that are not often asked but which belong to any such inquiry. First is the relevance of historical memory: Does collective memory have anything to do with the historical meaning of an event? Can we learn something about meaning from memory? Does the reconstruction of personal knowledge into collective memory offer any important clues to the historian? Second is the role of experience itself. Does experience at the Fair coincide with the intent of the managers? How did visitors behave? Did their daily itineraries reveal anything about their understanding of the Exposition? If patrons largely constructed their own meanings (and if we can know this), how does this disjunction affect our interpretation of the event? Or does it?

Among the self-advertised intentions of the Fair administration were multiple purposes: to justify American intervention in and conquest of the Philippines; to present a world schema of scientifically based racial and cultural hierarchies; to introduce Americans to all manner of new consumer goods and advanced technology; to promote St. Louis as a world-class city; to entertain millions with inoffensive but lucrative entertainments; and, finally, to demonstrate a system of thought and classification that could comprehend everything constructive that humans had achieved up to 1904. But which of these, any, or all, were understood and experienced by those in attendance? Or perhaps better, how and in what contexts did visitors experience these ideas and demonstrations?

Before we can respond to these latter challenges, we need to examine what historians have written to understand how a dominant narrative has emerged. The first comprehensive history of the Louisiana Purchase Exposition was authored by Mark Bennett, manager of the Division of Press and Publicity, and published by the Universal Exposition Publishing Company with the clear approval of the Fair administration (the secretary of the Louisiana Purchase Exposition Company, Walter Stevens, contributed the introduction). Bennett obviously had access to company documents and scrapbooks. This makes his work a valuable compendium of management attitudes and a clear indication (after the fact) of what they hoped they had accomplished.

In his forward to the book, Stevens called expositions the "time-keepers of progress." "New wants were born in millions of minds," he wrote approvingly of the impetus to consumption provided by the display of new products. And he repeated the slogan, process and products, as the theme of the Fair. In the huge and comprehensive work that followed, Bennett tried to demonstrate Stevens's declaration that the Exposition was "an Encyclopedia of complete and accurate Data of the World's Progress." The work was crammed with impressive figures about huge crowds, popular displays, celebrity visits, and, of course, the Philippine exhibit, which, he noted, ran from "the lowest types of head-hunting savagery to the best products of Christian civilization and culture." In the midst of this celebration, Bennett also noted that about 40 percent of the admissions came from residents of St. Louis itself, but he did not then recalculate or estimate the real number of different individuals who attended the Fair. With an extensive photographic program, the author managed to convey a sense of thorough representation. In effect, this first history provides a significant historical source but is noteworthy primarily for its self-congratulatory interpretation, and for suggesting how the leaders of the Fair wished it to be understood in the future.[35]

For a long period, however, historians paid far more attention to other American expositions, and some even judged the St. Louis event to be a subject of marginal interest by comparison. Prior to Robert Rydell's redefining work on American world's fairs in 1984 (*All the World's a Fair: Visions of Empire at American International Expositions, 1876–1916*), historians engaged St. Louis as a peripheral event, evaluating the Fair as a minor occurrence of primarily local or specialized significance. A. W. Coats, in his 1961 article on the intellectual and scientific congresses held in St. Louis in 1904, suggested that the moderate intellectual success stood out amidst the din and confusion of the rest of the event, for example. Perhaps the least positive evaluation of St. Louis appeared in an ungenerous entry in the 1972 *Encyclopedia Britannica*, written by Frederick Pittera. As he put it, "Culturally and aesthetically its influence was slight, and it was classed as a dismal financial failure." While he acknowledged the unprecedented physical size of the grounds and the amount of exhibition space—and even the display of technological progress since Chicago's White City of 1893—he determined that St. Louis fell short of that earlier exposition in attendance and ran a serious financial deficit.

This contrast suggests the persistence of competition between Chicago and St. Louis that prevailed at the turn of the century and that is sometimes stressed in subsequent historical works comparing the two Fairs. Inevitably, there has been the occasional note of defensiveness in historical accounts asserting that serious attention should be paid to the Louisiana Purchase Exposition.[36]

Recent American historians, however, have expressed far greater interest in the extensive and rich archival record left by the Fair administration, to the printed sources influenced by the enormous publicity effort, and to accounts by well-known visitors such as Henry Adams. The advantage of this examination is that it allows for a critical exploration of the intentions of the planners. It emphasizes what they believed they accomplished, as reported in their accounts of the Fair's purposes and successes. At the same time, historians by and large ignore memory and the institutions that preserve memory. Nor do they generally attempt to explore the experience of the millions of Americans as well as foreign tourists who visited the Fair. In its most recent iterations, historical literature has been highly critical of the racial assumptions and imperial ambitions of many of those connected with the Exposition. Increasingly, there is a consensus that considers the Fair within a limited set of historical contexts: contemporary world's fairs in Chicago, Paris, and elsewhere; the imperial struggles among Western nations at the turn of the century; a crucial moment in urban history with an emphasis on "City Beautiful" planning and modern attitudes toward the city; and a dramatic exposition of American attitudes toward race and ethnicity. In particular, this last issue has also been defined in the context of the history and development of professional anthropology in the United States. And all of these interpretations are premised on the unique opportunity that fairs seem to present as dramatic moments of introspection and as opportunities for the dissemination of new ideas. An unspoken assumption is that the ideas promoted by the designers of a fair were legible to the tourists who attended. The historical meanings of the fairs were thus noteworthy because they fulfilled their expressed purposes.

James Neal Primm's biography of the city, *Lion of the Valley: St. Louis, Missouri, 1764–1980*, was the first modern historical reevaluation of the Fair, and Primm told a positive story of the Exposition, but in the context of city's history. Published first in 1981 and then revised and republished in 1998, the work takes a comprehensive, long-term, wide-angle view of the Fair, judging it to be a principal episode in city building. This largely institutional and political history stresses the organization of the Fair, the biographies of its leaders, and the permanent changes wrought upon the surface and infrastructure of the city. While describing a few of the exhibits, particularly on the Pike and at the Philippine Reservation, Primm evaluates the Fair according to its attendance and finances, both of which he estimates to be very positive, although he notes that the total figures included workers' entries and exits. Unlike the example of Chicago in 1893, he writes, more than half of the visitors came from outside of the city. And beyond its intellectual congresses, the Fair was noteworthy as the venue for the Democratic National Convention of 1904 and the first site of the modern Olympics in the United States.[37] Significant permanent changes in the city resulted from the enormous effort to put on the show: improvements in the water system, urban transportation, and the condition of streets and the construction of new hotels. Even if only a few permanent structures from the Fair survived, such as the Palace of Fine Arts (which eventually housed the St. Louis Art Museum) and several state pavilions that remained as private residences, the Fair indelibly marked the history of the city as a brilliant and colorful moment and a high point in its history. What should be noted about this work is that Primm's description was one of the few historical works that bridged the gap between historical writing and celebratory versions of collective memory frequently invoked throughout the twentieth century as the city feted a history that looked considerably brighter than its present or future as time passed.[38]

One of the first works to remark on the important issue of anthropology at world's fairs was a book of essays published in 1983 in conjunction with an exhibition of photographs and artifacts of the San Francisco Exposition of 1915. Anthropologist Burton Benedict devoted the introductory essay to this volume to anthropological exhibits at San Francisco and St. Louis eleven years earlier. The anthropological theories and assumptions that structured the exhibits at both venues constitute his target. The result is a fascinating comparative reading, because it focuses on the structures and ideas repeated at several fairs included in anthropological exhibits, elaborated in classification systems, and discussed at intellectual congresses. Perhaps most significant about this piece is the attempt to describe and periodize a developing world's fair culture typical for this age. Benedict constructs a visible line of development in modes of consumption on display in these events, from what he calls an initial profusion of decorative manufactures laid out for a new middle class to later fairs that displayed social amenities and services for a wider segment of the population. He also provides a classification vernacular for the large variety of human subjects and types on display in one form or another at the fairs and invents a word to designate each sort. These included, he notes, technicians, craftsmen, freaks, trophies, and specimens. The Philippine Reservation at St. Louis, he concludes, was an example of trophy display: showing the fruits of a successful war with Spain and the subsequent conquest of the Philippine Archipelago.

If fairs constructed hierarchies of taste by positioning goods and prizes, that is because they were, he notes, celebrations of a world constructed by humans. Classified and ordered, these objects could be giganticized or miniaturized for the sake of display. Thus, he concludes, scale models demonstrated the ability of modern society to reduce an object to a workable, consumable size, while oversized display pavilions publicized the equally powerful ability of industrial society to contain all the objects it produced, no matter what their dimension or function. This early culture of fairs eventually disappeared, he writes, as the purposes of international exhibits evolved and as general, integrated displays were replaced by national and corporate exhibits. Although all world's fairs could be understood as rituals of competition, he concludes, the terms of this interaction evolved to reflect changing modes of consumption.[39]

A major variation on this theme of the cultural significance of world's fair culture can be found in the writings of Neil Harris. Although Harris has published his ideas piecemeal, sometimes in journals generally unfamiliar to historians, his 1990 collection, *Cultural Excursions: Marketing Appetites and Cultural Tastes in Modern America,* made a clear and convincing case for recognizing the relationship among various modern cultural institutions such as department stores, museums, and world's fairs, all contributing to an emerging culture of display that typified modernizing America. This notion, while not focusing on anthropology or imperialism or racism, proposed assessing fairs in their larger cultural contexts, something that other historians have emphasized far less. By introducing the extended contexts in which fairs existed, Harris recognized their shared interactions with other, similar cultural events of the day. If this diminished their singularity, it argued that they belonged to a general culture of modernization.[40]

The most influential recent works on the St. Louis Exposition have been written by Robert Rydell; indeed, he revolutionized the way historians understand late-Victorian world's fairs. This work was initiated in a 1980 dissertation, followed by two preliminary articles appearing in 1981 and 1983, before the publication of his first important book in 1984, *All the World's a Fair; Visions of Empire at American International Exhibitions, 1876–1916.* Considering the major fairs held between 1876 in Philadelphia and 1915 in San Francisco as a unified subject, Rydell developed several points that have become standard interpretations of the principal expositions during that era. His argument stresses two interlocking ideological categories that he believes structured these fairs. First is the discourse of a racialized anthropology, accompanying the explication and justification of America's late-century imperial aspirations. All of these fairs, he contends, depended on the same display of motifs, although he singles out St. Louis for the enormous size and thoroughness of its demonstrations. The Philippine Reservation, officially sponsored and supported by the U.S. Philippine Commission, with its village types and model school, constituted a key integer in this equation, something underscored by the significant participation of leading American anthropologists like Franz Boas and Ales Hrdlicka along with W. J. McGee, the leading force behind the 1904 exhibits. As he paraphrases it, McGee's anthropological thinking constituted a particular interpretation of the "white man's burden" appended to an evolutionary scheme that divided the stages of human development into four groups, rising from savagery to barbarism, to civilization, and finally to enlightenment. He also suggests that the Philippine exhibit loomed very large in visitor interest, and to prove this point, he quotes David Francis to the effect that 99 of 100 fair visitors made their way to the Philippine Reservation.

Comparing the cost of most other large displays at the Fair, Rydell also finds that the investment in the Philippine exhibit was significantly higher. Of the $8,000,000 or so spent by all foreign countries to mount exhibits, France spent $1,500,000 (including colonial exhibits), and Germany spent slightly less, with Great Britain in third place. The total cost of the Philippine Reservation, on the other hand, was $1,500,000, which is as much as any single one of these foreign nations expended.[41]

Rydell's second argument, a variant of the racialized division of the world into colonies and empires, constitutes what he calls the "utopian dimension to American imperialism." This visionary policy suggests that a dream of social and economic melioration underlay the categories of assertions of superiority and force. Just as the French had their *mission civilatrice* (a presumption that French colonization and empire would bring light and progress to places otherwise condemned to perpetual darkness), so America's Indian schools and, based upon them, a benevolent imperialism in the Philippines would bring enlightenment to all but the most marginal groups—like the dark-skinned Negritos, who seemed destined for extinction. This rationalization imagined that the much debated entrance of the United States into competition for empire aided our new subjects as much as America gained, even if Washington appeared to be offering civilization as a fragile veneer to its ulterior motives. To Rydell, even the Pike appeared to reflect this dual educational purpose. In the context of amusements and concessions depicting exotic cultures, imperialism and racial categorization "could be fun." As he summarizes it, these Victorian Era fairs reflected "an organizing principle of American Life." The fairs provided a visual and palatable object lesson in the virtues and rewards of social order at home, interpreted and justified by an up-to-date evolutionary anthropology and sociology. They imparted a powerful argument for the special role America had been chosen to play during this last act of the scramble for empire. By linking the conquest of the American continent and its native people through the commemorations of Columbus in 1893 and the Louisiana Purchase in 1904, these fairs revised American history to fit a larger contemporary purpose, which was a promise of orderly industrial progress internally and a world newly opened to American power and civilization.[42]

The energy and acuity of Rydell's critique in these works comes from his rejection of nineteenth-century racial and colonial assumptions. It derives its considerable moral and critical force by tapping into the present-day revolution in attitudes toward race in America. This is as it should be. Historians often fruitfully explore the disjunction between their own beliefs and the social assumptions of the past to generate distinctions and dissonances that require historical investigation. The divergence of moral assumptions opens the door wide to interpretation; change makes difference visible and puzzling. In this sense, history radically differs from collective or even individual memory, one of whose tasks is to maintain continuity with the past, not to dwell upon rupture and distinction and reevaluation.

Rydell also makes an important methodological choice to stress the centrality of international expositions as educational forums. This accords a particular importance to the role of specific key individuals and their elaborations of the same principal points about race and imperialism that appeared over and over again. Although he does cite a collection of transcribed memories of the St. Louis Exposition, principally to confirm the importance of the Philippine exhibit, he is not, by and large, concerned with what visitors might have seen or done at the Fair. The racial and imperial discourses as he presents them are univocal and unidirectional. They infect the actions, speeches, and correspondence of leading individuals associated with the fair administrations and, by implication, contemporary audiences. This interpretation is further supported by an astute reading of the geography of the exhibits, explaining the hierarchies of placement validated by the immense and telling photographic documentation of these events. The fairs, he concludes, constituted an important one-way dialogue that both stimulated and illustrated American culture at a pivot point. They can be read, in other words, as documents that reveal significant truths about American culture and ideology, organized and imposed from the top down.

As the current master narrative in the historiography of the St. Louis Fair as well as several contemporaneous exhibitions and documentary films,

Rydell's books occupy the center space of exposition history. At the edges circulate a number of other books and articles published since 1984 that extend his insights or emphasize a slightly different element of Fair history but do not take issue with the fundamental point. One of the most inventive and persuasive of these is Beverly Grindstaff's article "Creating Identity: Exhibiting the Philippines at the 1904 Louisiana Purchase Exposition," published in 1999. Accompanying an imaginary fairgoer on a visit through the Philippine compound, she envisions the experience of confronting the carefully placed racial and ethnic messages. Taken together, these messages, she concludes, excluded any political discourse except the necessity of American stewardship over the islands. Instead of a Filipino national identity, the exhibit racialized anthropological differences, transforming people into objects of science rather than citizens of a nation. In effect, this is an essay on experience but imagined through the intentions of the anthropologists.[43]

If interpretations such as this have, by and large, prevailed among historians, they have also introduced two important assumptions into the mainstream of interpretation, where they now reside as givens. The first is the singular importance of world's fairs as stimulants of ideas and ideology. The second assumes that the visitors understood and absorbed the messages promoted by the fair managers and the anthropologists who worked to design the exhibits. But only by exploring the experience of patrons at the Fair can we determine whether or not either of these hypotheses is valid.

The devastating and distorted impact of anthropology, racism, and Social Darwinist visions of evolution identified by Rydell can be illustrated in their individual consequences in the compelling biography of Ota Benga, a Pygmy brought to St. Louis as part of the anthropological display. Only four feet eight inches tall, with filed front teeth, and taken from the Belgian Congo area, Benga fit into McGee's agenda to assemble "representatives of all the world's races, ranging from smallest pygmies to the most gigantic peoples, from the darkest blacks to the dominant whites." The biographers describe Benga's peripatetic, short, and tragic life after St. Louis. He first traveled back to Africa, then returned to the United States for a temporary stay in New York. There, Benga was first on display at the Museum of Natural History, then, for a grotesque moment, a resident at the Bronx Zoo. Finally, he came to reside in a black community in Lynchburg, Virginia, where he committed suicide in 1916. The story of this ruptured and discarded life epitomizes the damage and cruelties inevitable in the treatment of humans as anthropological specimens. Once culled from the forest, Benga could not return, nor could he find a place as an object of display, nor eventually even in the community of African Americans in Lynchburg that sought to welcome him.[44]

As an exemplary story, the sad fate of Benga is also an indictment of American racial practices and assumptions, as well as a chilling display of careless ignorance and manipulation by some of the leading anthropologists of the day. The authors do not limit themselves to biography, but write this story as one act within the Fair's other racial dramas, including the Boer War reenactment of battles against Zulus in South Africa and the various controversies that surfaced around Benga and other native peoples on display. The book reproduces newspaper accounts of persistent attempts to swindle the Pygmies. Fairgoers apparently tried to defraud them for the privilege of a photograph by tendering steel discs rather than coins in payment, thinking that the native peoples would not understand the meaning of money. Other articles quoted in the book had such lurid titles as "Pygmy Dance Starts Panic in Fair Plaza" and appeared in the *St. Louis Post-Dispatch,* one of the city's most reputable papers. Another article in the same paper noted an incident where "Enraged Pygmies Attack Visitor." Seen in the context of the misunderstandings, racial stereotype, and journalistic exploitation, the tragic outcome of Benga's life cannot be too surprising. For the authors, the presentation of anthropology at the Fair was a sad miscalculation with fatal consequences. It provided a lesson in confused moral priorities that provoked an inevitable tragedy.[45]

If the story of Ota Benga is one of misfortune made inevitable by misguided science and racism, inflicted on a hapless individual, it was not entirely clear to some authors that the science of humanity actually spoke with one voice at St. Louis. As John Troutman and Nancy Parezo argue in their 1998 article for *Museum Anthropology,* one of the motivations of anthropologists at all of these contemporary international expositions was to establish the field itself against competing narratives of religious leaders, missionaries, and entertainment entrepreneurs who had a differing, if equally strong, interest in using subjected and colonial peoples. While McGee sought to popularize his four-stage cultural development scheme as the best way to understand the distinctions between savagery and civilization, Native Americans were depicted as an exception in this schema. They were portrayed as a dying race, a vanishing people whose study presented an immediate opportunity—even an emergency.[46] Looking carefully at the ideas surrounding their display, Troutman and Parezo note that McGee and other anthropologists actually articulated two hierarchical and sometimes incompatible schemes, one based on culture (savagery developing toward civilization) and the other based on race. This, they maintain, sent a "clouded message to the media," in particular. This contradiction was especially evident in the depiction of Native Americans as capable of elevation "to the ideas of Western Civilization," a notion that, by allowing for progress and change, would certainly subvert notions of fixed racial hierarchies. In the end, the authors suggest, such distinctions and arguments about race might not have mattered much anyway. Interviews with surviving visitors to the Fair, they indicate, generally contained no mention of the racial ideologies that motivated anthropologists. Perhaps, the authors speculate, visitors "did not know how to or did not wish to reflect on the exhibit's messages."[47]

Paul Kramer, in his pioneering article for *Radical History Review* and recent expanded chapter in *The Blood of Government: Race, Empire, the United States and the Philippines,* has found a complex significance to the anthropological exhibits of St. Louis reflected in a series of mixed messages and contradictory outcomes. For one, he notes, the creators of the Philippine Reservation themselves concluded that the project was, ultimately, a failure, and they did not attempt to remount it at subsequent fairs. Moreover, he suggests that the essential ideology of race was itself contradictory, and this distorted the clarity of the displays and their messages.

Kramer argues that confusion began with the multiple objectives of the display and the competing intentions of various players, including vice president of the United States and former governor of the islands William Taft, the Fair administration, anthropologists, journalists and audiences, and the Filipinos themselves, all of whom, through their interactions, helped shape the experience. The essential problem was the contradiction between notions of primitivism and hopes for uplift, between a racial ideology that judged people as inferior and an ideology that justified imperialism through its potential to civilize and change them. As Kramer concludes, this forced the Exposition planners to "walk a tightrope between racist hierarchies of the Filipinos' inferiority and paternalistic promises."[48]

This deep contradiction played out in a variety of specific ways. There was the horrified and ferocious local reaction to dating between Filipino men and white St. Louis women, with violence over this issue erupting between U.S. Marines and the Filipino Scouts. There were gawking audiences at the Igorot Village attracted to lurid stories of their strange appetites and male nudity. Many Filipinos, Kramer notes, refused to participate in the racial dramas staged by anthropologists and by the U.S. Philippine Commission. In the end, the contending purposes and the many actors involved in this racial production prevented any single message from predominating.[49]

The most recent and comprehensive exploration of the racial displays at the Fair finds unity, not contradiction, in the presentation and effects. Indeed, in their comprehensive *Anthropology Goes to the Fair* (2007), anthropologists Nancy Parezo and Don Flower argue, "McGee's version of science would be the central theme for the entire exposition, seen by millions of people and carried away in the form of racial stereotypes." Considering the large Native American exhibits, Indian school, Philippine Reservation, and countless other anthropological venues, they assert that the interaction between many of the elements of the Fair created a total ambiance of highly charged racial attitudes. The moment also represented, they conclude, the beginning of a turning point, when anthropology began to backtrack against theories of racial categories and stages of civilization.[50]

In his 1997 book, William Everdell assembles and integrates a number of the interpretive points discussed thus far. He places them in juxtaposition and defines this state of integrated difference as typical of American modernism. Choosing "Meet Me in St. Louis" as his epigraph/title and opening of the first paragraph, he means literally in this book to discover the odd coincidences, the accidental encounters of people and culture that the Fair contrived to bring together. The city of St. Louis during the summer of 1904 and the Fair itself contained the jumbled cultural elements of modernity, he contends. Beginning by quoting Henry Adams's nostalgic recitative on order in the modern and medieval worlds from *The Education of Henry Adams*, he juxtaposes this lament with the "invention" of the ice cream cone, the Ferris wheel, and Jim Key, "the educated horse" and star of the Pike. He then moves on to consider music in a celebration of St. Louis's contribution to popular culture—particularly the ragtime compositions and performances by Scott Joplin. However Joplin himself was treated or mistreated at the Fair—whether he could dine at its eateries or move around unmolested—his music was becoming a celebrated element of contemporary popular culture. Nonetheless, it was also clear that other, negative racial dramas were at play, particularly in the assumptions about ethnicity and race exhibited by the Olympic Games and the mocking, shadow contest called the Anthropology Days, during which the Fair's exotic peoples competed for athletic prizes, with representatives of industrial countries absent.

Everdell describes the international congresses held on the fairgrounds as the intellectual counterpart to this modernist collage of juxtapositions. In field after field, he writes, it appeared that scientists and intellectuals were on the verge of articulating the major ideas that would eventually dominate the twentieth century. Even if Freud and Einstein were absent, their ideas stood in the wings, anticipated by those who had neither the genius nor inclination to articulate them. Taken as a whole, the point of this assessment is to understand that St. Louis represented both a moment and a rendezvous: a paradigm of modernism, an anticipation of change, because it attracted so many new ideas and attractions and visitors to see the world as it was becoming but had not yet become.[51]

One final work in this large and inventive historiography is Eric Breitbart's, *A World on Display, 1904: Photographs from the St. Louis World's Fair*, published in 1997. This thoughtful consideration of images of the Fair, primarily photographed by seven men and women, creates a visual counterpoint to the argument that anthropology and fascination with racial types and generic faces preoccupied visitors. Although most historians of the Fair use its rich photo resources from 1904 to illustrate their arguments, Breitbart places the image foremost, making his subject matter the style, the attitudes of the photographer, and the captured image. While this account does not substantially change the main argument about the prominence of anthropology, nor the racial distinctions and divisions and categories that marked it, his strategy suggests the importance of a visual counterpoint to arguments developed out of literary sources. At the same time, he suggests, much of this photography is specialized, ranging from carefully framed studio portraits to more candid shots. In other words, these photographs reproduce images and scenes that visitors might have seen and experienced, although not others that were taken in the posed settings that professional photographers often favored. Breitbart concludes from his study that the photograph served to transform people and objects into items available for consumption. This transaction, he notes, often conveyed the ideology of the Fair promoters and anthropologists to the purchaser, who might not have attended the fair.[52]

Although it is rare for historians to achieve consensus regarding a momentous event, the Louisiana Purchase Exposition seems to be one such exception. Of course, some authors differ in emphasis, primarily because of the way they contextualize their arguments, some focusing on one element, such as the intellectual congresses, and others on the Philippines, and still others on the importance of city planning and administration. But taken together the picture that has emerged is remarkably unmuddied by disagreement. Several general conclusions appear to be in place.

1. The St. Louis World's Fair was a significant national event that illustrates in a powerful and legible fashion the prevailing and emerging culture of the day. Because it illustrated and instructed its visitors, the Fair provides us with an unusual window into a dynamic cultural moment in American history.

2. The prevailing ideology of the exhibition, evident throughout its displays, was an evolutionary scheme of peoples and cultures that placed Western civilization and the United States at its pinnacle and subjugated colonial populations on the lower rungs. This evolution narrative is capacious enough to include the history of the United States itself, the Purchase of the Louisiana Territory, and the suppression and removal of its native peoples to reservations or oblivion. Although historians may stress elements in this

general narrative more or less, the implications of the story are summarized in America's imperial ambitions displayed in the Philippine Reservation of reconstructed villages, in the reenactment of the Battle for Manila Bay with the Spanish, and in the Wild West shows and collections of Native Americans on the Indian Reservation.

3. A third generalization is that the St. Louis Exposition exemplified a burgeoning consumer culture, apparent everywhere in a nation emerging from the regimen of necessity. This transformation rendered everything—from the essential to the luxurious, from history to religion, from experience to memory itself—into possible objects of consumption and spectacle. Although this point appears somewhat less prominent in some works, it is almost universally present in some form.

4. There is also the question of St. Louis itself—the fourth largest American city at that moment but one destined to slip into a less prominent role as a minor regional power in comparison to Chicago. For historians, the Fair is an oversized event in the story of an otherwise unremarkable American city—the most important event, perhaps, of its history. While the Fair clearly transformed St. Louis as an urban environment, joining it with other forward-looking midwestern cities like Chicago striving to create a Beaux-Arts embodiment of the City Beautiful movement in its public buildings, it became an event that figured ever larger in the history of the city as the fortunes of the city itself diminished.

5. Finally, historians and memory custodians both have proclaimed the centrality of modernism expressed in the St. Louis Exposition. The Fair, as David Francis argued, was a summation of the latest advances of civilization, and later commentators have pretty much agreed. It is certainly true that many of the latest advances in technology were on display: a potpourri of inventions and consumer items like automobiles, huge power generators, advanced locomotives, telephones, motion pictures, fingerprinting, and so on. These items would, in the future, contribute mightily to the new shape of modern society. As Henry Adams sarcastically put it, "a third-rate town" with no "history, education, unity, or art" had put on a magnificent show. "The world had never witnessed so marvelous a phantasm." To Adams, this remarkable Exposition revealed the organizing power of modern technology, the epiphenomena of the organizing energy of the dynamo.[53] Despite his ironies and snobbery, Adams was saying what later observers would repeat: the St. Louis Fair was both the symbol and enactment of a certain mode of modernism.

But what do we mean by *modernism* when referring to the Fair? To what degree did the Fair embody the complexities we identify as modernism? We must hesitate before embracing such descriptions, for there was much unseen and unheard and definitive about modernity that did not find its way onto the fairgrounds in St. Louis. A good example can be found in the styles and uses of architecture. If we define modernity in 1904 as the accumulating possibilities of new building materials and structural designs, we find two broad possibilities. One stressed the visual preeminence of these materials and structures. The other emphasized the way imaginative façades and surfaces could be placed on these structures. While St. Louis itself had erected several splendid examples of advanced architecture, particularly in its downtown skyscrapers, the Fair presented a cloying and thrice-derived Beaux-Arts design, developed largely in France and reproduced at both the Chicago and Buffalo fairs. What *is* modern about these huge palaces is surely not their exterior decorations, but rather the expression of architectural potential: the possibility of using new pliable materials to embody fantasy and imagination. The technical skill for rapidly constructing a large framework that could be covered in new sorts of plastic materials certainly represents an essential component of modernism in architecture. But the look of the exhibit palaces of the utopian city constructed at St. Louis was stubbornly archaic. Thus, the insecurities of American urban ideals, coupled with the ability to erect a fanciful celebration of the architectural past once removed by geography and then again by history, gave the Fair its fundamental eclectic character.

Modernity may have been an intellectual theme at the Fair, but some of its principal representations were absent, particularly from representations of culture. While it is true that ragtime on the grounds represented an important suggestion of the direction of popular music, impressionist and dissonant music and Postimpressionist painting were absent. Perhaps most important, the celebration of imperialism, with its conquered Native Americans, Filipinos, and other colonial peoples, took no notice of growing opposition to this world order—or even opposition within the United States, despite W. E. B. Du Bois's *Souls of Black Folk,* which had been published in 1903. The reenactments of the Boer War and the naval battles around the Spanish-American War took no cognizance of what would be major themes of the twentieth century: anti-imperialism and nationalism, any more than the Philippine exhibit recognized the powerful guerilla war mounted against the American seizure of the Spanish colony and still sputtering on.

Despite Hugo Munsterberg's determination that hierarchy and order could be imposed on all fields of knowledge, the international intellectual congress summed up, at best, an intellectual world that would shortly face challenge and disarray. Instead, the possibilities of modern organization seemed more apparent in the extraordinary feats of organization and publicity undertaken by the Louisiana Purchase Exposition Company. Although many of the displays featured examples of advanced technology and invention, their meaning and impact had not yet become apparent at the Fair and may only seem so in retrospect. As I will argue in the chapter on experience, visitors rarely reacted to them or understood them in the context of their own lives and occupations. Most significant, some of the impressive technological achievements, such as the Ferris wheel (already old by 1904), the baby incubator (displayed also at Coney Island), and the nighttime lighting, were employed primarily for amusement purposes. Likewise, the most mentioned "inventions" of the Fair were consumer items like hot dogs, ice cream cones, and iced tea. It is, perhaps, in this form of consumption and entertainment that modernity was most visible on the fairgrounds.

If historians are, by inclination, predisposed to step carefully around the pitfalls of nostalgia, at the same time, nostalgia and memory persistently raise hesitations about these historical constructions, registering a strong caveat about the critical depictions that constitute accepted narratives. In fact, historians have, by and large, ignored memory of the Fair in whatever guise it has been expressed, relying instead on printed sources, the archives and correspondence of leaders, and officially generated or approved photographs of the event, as well as impressions of well-known visitors. What remains generally absent from this work is the individual viewpoint (expressed singularly or as part of a pattern of behavior), the testimony of the consumers who were not just the object of the Fair, but also the subject of the social transformation we call the consumer revolution. This is all the more reason to explore what they consumed.

I have insisted, in this book, that historians can learn from memory, can perhaps employ it to see through to contemporary experience. Certainly there are good reasons to be cautious in opening such a line of inquiry. Many years ago, Warren Susman raised the question of memory in connection with research he was doing for a study of American expatriates in France during the 1920s. As he interviewed key participants, he was perplexed to discover that "frequently they knew less about their lives than I did from reading other sources." Their memories, he concluded, had been adulterated by reading the accounts of others, by stories in the media, by autobiographies and movies. Susman decided, therefore, that he had inadvertently stumbled upon the way the mass media had transformed (and inevitably transforms) experience into myth. Even more disconcerting, he found that his own reading and internalization of the stories he studied could on occasion become a sort of false memory engendering the feeling that he had been present at events in other people's lives. Susman's surprise and uneasiness has much to do with two fundamental considerations that make historians wary of memory: that it is frequently inexact and that it is corrupted by the superimposition of other accounts by other participants and in various forms of media, such as films, popular histories, reconstructions and historic exhibitions, and photographs.[54] Along with these, two other problems must be mentioned. The language of memory is often individual and rarely analytical or social. At the same time, individual recall often conforms to collective memory or agreed-upon ideas whose source is not the individual who relates them. Individuals repeat the nuances of mood and the evaluation of historical events prevailing at the time of recounting without realizing it. In retellings of the past, nostalgia can play a significant, even overpowering role; it becomes a memory of pastness itself. Nostalgia also contains elements of a serious critique of the present. If all these problems be so, how can we rethink the issues of memory to help enlighten the puzzles of historical analysis? How can we make memory relevant to analysis? Can we ultimately use memory to enlighten history?

CHAPTER THREE

MAKING MEMORIES

Historians encounter memory in two principal forms: through their interest in oral history as source material and as experts working as consultants for exhibits, documentary films, and other commemorations that speak to an audience whose expectations and interests are shaped by some form of collective recall. A fascinating illustration of the latter is Eric Breitbart's documentary film of the Louisiana Purchase Exposition, *A World on Display,* produced in 1994. In a striking and unusual juxtaposition, he places interviews with (then) elderly visitors to the St. Louis Fair alongside analyses by historians explaining the significance of the event. Yet these two parallel discourses almost never intersect, nor do the participants ever interact—that is, historians do not comment on these interviews or their possible meanings. If nothing else, this disjunction underscores the gulf that exists between the different functions of memory and history.

At the same time, none of the interviews wanders into analysis or engages what we might define as historical questions. One man mentions how much he learned from the Fair, what a wonderful experience it represented and how novel the displays were, but he does not explain what any of these implicit comparisons might mean. Two of the women merely provide testimony that they attended. One reveals that she visited as part of a group; the other, that she was too poor to afford most of the concessions. Two other participants mention the controversial Igorot Village. One recalled that he was too young to understand very much, but that he did learn about the other peoples of the world, their "talents and abilities." Only on rare occasions does the narrative of the documentary use this testimony to stress a historical point. Instead, it appears that memories of the Fair are generally at odds with historical analysis. Frequently they are vague, sometimes quixotic, and always idiosyncratic. Why then are they present if the stories are not relevant or informative? Certainly they provide a sense of authenticity to the film. Whatever their content, they are testimonies to pastness itself. Their address to the audience is to affirm that they were present, that memories do still exist. So they become part of the larger documentary aesthetic, and, like the film footage and photographs from the Fair, they lend verisimilitude to the project.[1]

The documentary also demonstrates that historians strongly disagree with commonly held collective memories about the meaning and significance of the Fair. This is often the case in this medium, for, indeed, it could seriously be argued that one purpose of the documentary film (as well as museum exhibitions) is to challenge and modify collective memory, to confront nostalgic meanings and bland conclusions commonly held. Aimed at popular audiences, documentaries place visual images such as photographs and old films, maps, paintings, and other memorabilia into new contexts and arguments. They recontextualize oral history. Their purpose is rupture and reconstitution, a re-creation of memory set in the context of a historical argument. The most powerful narrative in Breitbart's film is unmistakably the anthropological and racial exhibit. Its elucidation is a narrative thread that runs through the production. While Robert Rydell is the principal source of this orientation, two other experts, Zeynep Celik, an architectural historian, and the anthropologist Ted Jojoa, confirm and expand this interpretation. A second theme is the importance of an expanding consumer culture, which the group suggests is also a major historical transformation visible in St. Louis, a shift in cultural focus from production to consumption. This means, as Rydell explains, the transition of the average American from an occasional purchaser of necessary goods to a person whose life is defined around objects of consumption and the dreams of their acquisition.

Although briefly expressed, there is a strong, contending interpretation of the Fair suggested by historian Neil Harris. Rather than emphasizing racial and imperial dramas or modern economic patterns, Harris suggests something else. He first describes the importance of the St. Louis Fair as a model city, an idealized, functioning urban environment where every service and facility is organized and efficient. He then links this construct to the format of other contemporary world's fairs, all of which venture in this direction of urban Utopia. His second point is to remark upon the attractiveness of the Japanese exhibit and the aesthetic of integrating old and new, the traditional with the innovative, in a mixture that greatly influenced American design.

These efforts to challenge the collective recall of the Fair as a moment of happy memories of old St. Louis are sustained almost throughout the documentary. Only when credits roll at the end does the soundtrack permit nostalgia to break through as it plays an early recording of the song "Meet Me in St. Louis, Louis." But this is almost an afterthought.

Rather than see this documentary as an anomaly or in any sense a failure—which it clearly is not—I would stress the remarkable and vivid way that it unconsciously opposes history to memory, the way it articulates the problems of thinking through these two different formats. As Paul Ricoeur has written, "History would like to reduce memory to the status of one object among others in its field of investigation; on the other hand, collective memory opposes its resources of commemoration to the enterprise of neutralizing lived significance under the distant gaze of the historian."[2] In practical terms, the film raises a serious question that historians need to confront: Does memory really matter? Is it possible to isolate experience from the shroud of collective memory? Can history be truly convincing if it contradicts memory? For historians, is it significant whether or not visitors to the Fair perceived the messages that were articulated in the displays? Were patrons of the Fair present at an event whose significance was to them so individual and egocentric that it cannot be combined into anything more concrete or specific than nostalgia for the good old days? Is there any way, finally, for historians to use memory and experience constructively? Or is Pierre Nora correct when he said, "History is perpetually suspicious of memory, and its true mission is to suppress and destroy it"?[3] Conversely, can the custodians of memory convincingly incorporate history into their narratives? Can history become collective memory?[4]

This chapter will explore the development and significance of the memory culture of the St. Louis Exposition as it emerged after the Fair and as it operates in the contemporary world. In some respects, the city has an unusual stake in preserving the memory of this great event; it is indelibly written into a past that colors the present with pride, nostalgia, and sometimes disappointment. This chapter will examine the theoretical issues involved in memory studies and oral history. It then considers the history of St. Louis memory organizations and commemorations, treating them as contributors to what seems like a developing historiography, a kind of "memorography" of collective recall. Finally, it will suggest what historians might learn from examining how both oral history and collective memory develop and how they can be used to articulate new insights.

There are four theoretical divisions of memory culture that require some initial definition and attention. These include collective memory, individual memory, the organizations and instruments that mediate memory, and the physical objects associated with recall and recollection (souvenirs, collectibles, and museum pieces and objects of veneration). Historians have certainly paid considerable attention to distinguishing between memory and history and describing the relationship between the two, but not generally in the context of events such as world's fairs. Their conversation has generally related to important, even traumatic political issues such as the validation of the Holocaust, questions about the effects of World War I, and oral histories designed to recapture the distant experience of slavery and sometimes the more contemporary civil rights movement. A scholarly journal founded almost twenty years ago, *History and Memory*, is devoted primarily to this major concern about the Holocaust and other memories of World War II in Europe. More recently in the United States, issues of memory and history have converged in a national political debate about "cultural literacy"—that is, what usable analogies, principles, and heroes from the past should constitute the operational historical knowledge of ordinary Americans. This latter problem raises questions about how collective memory might function in contemporary society. Does it exist apart from its individual manifestations? Is collective memory actually myth by another name? Are history and memory constitutionally distinct and, if so, in what ways?[5]

For historian Jacques Le Goff, the problem of disentangling memory and history is a relatively contemporaneous one, emerging with the modern period and the dramatic rupture with the past in catastrophic events such as the French Revolution. Of memory, he notes, "its workings are usually unconscious," making it more subject to manipulation. At the same time, he continues, "the discipline of history nourishes memory in turn, and enters into the great dialectical process of memory and forgetting experienced by individuals and societies." Le Goff is particularly eager to define modernity and its effects upon the relations between history and memory. Implicit in modernity, he concludes, with its fundamental reorganization of knowledge, its new theories and revolutionary mechanical means of recording and preserving the past, is a much-changed, and often challenged, role for memory.

The historian also suggests the reality and power of collective memory, a very real and observable phenomenon evidenced in commemorative culture: the spoken word, rituals, and festivals, as well as in new forms visible in the archives of film and photography. It is a primary mental and psychological function, he says, that allows groups to make sense of their pasts. While history is intimately related to memory, it is also quite clearly a different mode of thinking, a higher form, he would have it, of contemplating the meaning of the past. Thus, he concludes, "Just as the past is not history but the object of history, so memory is not history, but one of its objects and an elementary level of its development."[6]

As Le Goff suggests, memory, history, and identity are clearly interrelated if, at the same time, different from each other and even competitive, interpretations of the self. This proposition is the starting point for John Gillis in his collection of essays entitled *Commemorations: The Politics of National Identity*. In his introductory essay, Gillis describes the difficulty of enunciating a stable and precise definition for words like *memory* and *identity*, in part because the two reflect and define each other. Commemoration, he notes, involves coordinating and integrating individual and group memories around, for example, sacrosanct and familiar events such as the American and French revolutions. These events became embedded in memory particularly during the nineteenth century, when highly selective and gendered versions of these experiences became powerful remembered narratives. The history of the twentieth century, however, suggests the gradual disintegration—or perhaps, reintegration—of memory into new contexts, as additional groups such as women and ethnic or racial minorities clamor to be included in the collective memory of the past. Thus, as collective identities change, so do collective memories. And, if the presence of memory is less powerful in the present, he argues, this is because its singularity has been challenged as older holidays and patriotic commemorations have lost their exclusive meanings and thus their power of attraction.[7]

In a useful recent survey, Barbara Misztal locates the contemporary origin of interest in collective memory in debates about Vichy France and the Holocaust. Like other authors writing on this subject, she suggests that individual memory is socially organized and that collective meaning is an individually held representation of the past defined by a group. While there are other forms of memory—cognitive skills, for example, such as riding a bicycle—it is memory of the past that she believes has been decisive in helping to structure modern society. At the same time, various forms of memory have undergone radical reorientation due to the decline of tradition, the ideology of modernism itself, and the constant reconstruction of memory by the mass media. This comprehensive survey, starting from the theories of Maurice Halbwachs and his notion of the "frameworks of social memory," to Eric Hobsbawm's description of the invention of tradition, articulates the most important distinctions between history and memory. In the end, she concludes, perhaps the most positive uses of memory can be to correct social injustices; in other words, revision of the remembered past can remove a powerful justification for present-day injustices.[8]

The proliferation of memory and identity issues informing contemporary political issues has also attracted American commentators. In particular, patriotism and national commemorations are subjects of hot debate. During the 1980s, for example, this seemed particularly relevant because of political disputes centered on the design of the Vietnam Veterans Memorial in Washington, D.C., and questions about cultural and historical literacy among school children. Defining cultural literacy as appreciation for a pantheon of American political and military figures, one group of public intellectuals questioned whether history classes in public schools taught proper gratitude for the sacrifices and achievements of national heroes.

By defining collective memory in this politically charged fashion, writers such as Lynne Cheney and Allan Bloom, for example, confirmed what many commentators and theorists had long asserted: that collective memory was often an important political instrument, a public discourse infused with officially sanctioned stories or, as historian John Bodnar put it, "an ideological system with special language, beliefs, symbols, and stories." While collective memory, he writes, is often engaged in promoting social control and patriotism, it also competes with "vernacular" memory associated with the alternative ways that small communities or groups remember. The dynamic interaction at all levels of memory, he concludes, motivates a continuing struggle between the official and the local.[9]

If the mystic chords of collective memory are played in nation-building and patriotic harmony, then in the culture wars of the recent period, pedagogy is its most disputed instrument. Critics, whatever their political ideology, have almost uniformly faulted American schools and history teaching, or the absence of history from the classroom, for this assumed lack of a desirable collective memory. In a study published by the *Journal of American History*, Michael Frisch tested two propositions fundamental to this argument. The first confirmed some familiarity with traditional heroes and patriotic accounts of American history among the students of his college classes. Then he analyzed the effectiveness of teaching a more diverse, complex, and critical narrative of American history. Did exposure to different stories significantly alter the standard content of collective memory, he wanted to know?

As an experiment, Frisch examined students before they took college-level history and then afterward to find the preliminary content of common memory and the sorts of shifts that occurred because of the classroom instruction. What he discovered, however, was a relatively unchanged and consistent catalog of American heroes, regardless of what other significant individuals might have been introduced to students in their studies. This surprising continuity suggested two striking possibilities: that classroom instruction actually changed collective memory very little, and that the transmission of ideas and impressions that constitute collective memory may be extremely complex and embedded in a variety of cultural issues that resist challenge. The tenacity of collective memory, he concluded, had not "weakened in the slightest." Indeed, for the historian, that was the problem.[10]

Below the radar of debates about the collective memory of national commemorative events and patriotic figures, and certainly less consequential than the dramatic questions raised by remembrance of catastrophic events like the Holocaust, or the controversial Enola Gay Exhibition at the Air and Space Museum in 1994 and 1995, the problems of collective and individual memory of the St. Louis World's Fair remain just as complex and problematic for historians.[11] And just as surely, they demand interrogation; just as certainly they are subject to similar problems of distortion and corruption. And just as surely, we need to look carefully at what we might salvage from them.

One persistent problem is the modification of memory due to the temptation to reconstruct experience based upon the use-purposes of remembering, for example, finding a particular significance to a story for inclusion in the narrative of one's life. Another is, of course, the splicing in of additional information about an event gleaned from secondary sources other than experience. Just as the past changes over time as historians consider it differently and in new contexts, so participants may recall events in changed contexts. Historians Roy Rosenzweig and David Thelen discovered that individuals they studied in a project about memory and history learned about the past in a variety of situations, from museums and historical sites, to gatherings with their families. Often the memories they derived from these sources became recontextualized into some expression of self-identification reflecting their region, class, race, family, and other close institutions. For many of Rosenzweig and Thelen's subjects, history was exclusively a living presence in their lives, not academic history, but the personally significant past as revealed by historical sites, family histories, and even the Bible. The past, the authors discovered, was often invoked as a justification to change (or not change) the present or anticipate the future. While their subjects distrusted aspects of the mass media presentations of the past, they tended to privilege eyewitness accounts. While their subjects sometimes doubted professional historians as well as the patriotic narratives of collective memory, they were most attracted to history as it was remembered in the context of family, with photographs and other visual materials as prompts, and reflecting stories of family interaction and trips to historic sites as their sources of information.[12]

Though not an exhaustive exploration of the subject, my discussion has touched many of the primary issues raised during the debates about memory and history. As George Lipsitz summarizes them in his book *Time Passages: Collective Memory and American Popular Culture,* they include the Holocaust and issues of commemoration and the construction of sites of commemoration, the rise of skepticism about the accuracy of oral history, debates over cultural literacy, postmodern scholarship that raises fundamental questions about the nature of representation, the rise of interest-group perspectives in writing history and defining collective memory, and the location of memory within presentations of popular culture.

This last point has taken on increased seriousness during the twentieth century. The past one hundred years has witnessed an astonishing transition to a consumer society in which traditional narratives are continuously being undermined and supplanted by mass media presentations. The past, as much as anything else, now appears to be entertainment.[13] Or, as David Lowenthal writes, "The technology and artistry of retrieving the past often fuel popular interest in bygone times to the detriment of historical understanding."[14] Films, historic reenactments, documentaries, exhibitions, and historical fiction are the tools of the reproduction of collective memory. With a vastly increasing visual library of past images, and the possibility of their electronic manipulation, this tendency is likely to increase.

Evaluating the role of modern mass culture in remembrance, Alison Landsberg summons the striking medical metaphor "prosthetic memory" to indicate the palpable mental trace of an absent experience. As she argues, "a commodified mass culture opens up the possibility that people who share little in the way of cultural or ethnic background might come to share certain memories." In particular, she notes, film offers empathetic experiences like this. "Through mass culture," she continues, "people have the opportunity to enter into a relationship with the 'foreigner.' Gradually they learn to feel emotionally connected to what is intellectually a great remove." The author concludes that rapport stimulated by mass culture can be a means whereby different individuals acquire similar memories even if their experience includes no such common ground. These "imagined communities" of audiences provide a mutually shared imagery of identification. Thus, for example, the Holocaust has become almost a tangible American memory, with key commemorative sites such as the Holocaust museums established in several American cities. The extended television reenactment *Roots* in 1977 brought slavery closer to the nation's common heritage, and the title, itself, came to stand for the common national passion for researching family genealogies. Although Landsberg differs from other authors who have written pessimistically about the distortions of power and manipulation in the acquisition of collective memory, she is fully in agreement that the means of this acquisition are modern and that the imaginary power behind the construction of memory is the decisive factor.[15] Memory, under these circumstances, does not require experience of an event; it often recalls a past that was never experienced. It may be only a fragment left by the work of remembering what happened to someone else.

If there is one universal conclusion entertained by most theorists of collective memory, it is that memory itself is never a candid snapshot of the past, but always framed by culture and viewed through mutable lenses. Furthermore, it waxes with the push and pull of politics. It manifests in various social frames of reference: in families, groups, regions, ethnic groups, generational cohorts; indeed, in any form of individual or collective identity. Its ability to simultaneously maintain multiple valences suggests the social and political significance of constructing a usable past.[16] At the same time, to be usable, it must conform to the contemporary interests and needs of the individual and group.

Historians' access to memory often begins with oral history, a form of inquiry and record-keeping that recounts historical events in terms of individual participation, impressions, or recall. This mode of inquiry taps into conversations about the past but subjects the individual to informed questioning. Of course, it remains subject to the same distortions and inventions as collective memory. And individual recall gathered by oral historians also exhibits the added problem (as well as advantage) of often being a rehearsed narrative about individual experience, or "personal episodic memories." Often, these memories are accompanied by perceptible emotional states or significant cross-references. Sometimes they are obviously situated as parts of family memories in which various members play particular established roles in the recounting, so that the story affirms the "character" of the teller. Or they can be embedded in a generational identity. So too, they can be constructed to satisfy some sort of self-serving motive, or even reworked to fit contemporary opinions, or, finally, fashioned to please the questioner.[17] Does this mean that they are somehow always false or even useless and misleading? Not necessarily, but, like any other historical source, they must be subjected to scrutiny. As Alessandro Portelli writes, "memories of the past are, like all common-sense forms, strangely composite constructions, resembling a kind of geology." Digging down through these layers of meaning is thus the task of oral history.[18] As Paul Ricoeur concludes, everything starts "not from the archives, but from testimony." Perhaps this is why oral history and other forms of memoir lend gravitas and realism to history, allowing us "to assure ourselves that something did happen in the past."[19] And this, even when we know the perils of listening.

This theoretical work suggests several strategies for understanding the role of memory in our examination of the Louisiana Purchase Exposition of 1904 in St. Louis. The Fair occurred at a decisive moment in what historians of memory have designated as the shift to modernism, a change in the available means of producing memory. With the appearance of new instruments of mass culture, such as film, and a shift in the relationship of Americans to a developing consumer culture, the past itself was becoming a fungible item of consumption. The Fair was filled with anticipations of this new technology, from movies to automobiles to primitive airplanes—all of them in their fashion instruments of modern communication. The world of collapsing cultural difference made visible by imperialism was also very much on display in terms of the grand pavilions of foreign nations and the gathered peoples of colonialism. If, as theorists assert, memory is constructed and reconstructed, then the Fair provides a site of considerable interest to explore this process—and especially, how the Fair was established as a particular sort of event in the collective recall of St. Louis. Finally, this exploration should lead to suggestions about how historians might use memory to reassess their understanding of the Exposition.

The remainder of this chapter is divided into three portions. The first explores several ways in which individual memories of the Fair have been constructed; the second evaluates the role of institutions, organizations, and celebrations in establishing and maintaining a collective memory; and the third briefly indicates how individual memory might be used to develop alternative interpretations of the Fair.

In my discussion, I will define individual memory in the following fashion: It is *anticipation* joined to *retrospection* by means of an experience or an imagined experience. Initially, experience may be dominant, but over time and space, memory comes to depend more and more upon imagination. As it does so, it increasingly becomes subject to the emendations of collective memory. Individual memory cannot exist for long, if ever, in isolation, for it is constantly subject to review and reconstruction, if only through retelling or recollection. Yet it is still not exactly the same as collective memory, even if it reproduces elements drawn from that broader form of discourse. Individual memory may be made up entirely of reconstructed elements; it may reproduce the prevailing social and ideological interpretations of the past, yet it almost always situates the individual witness at the center of experience. In this sense, the idiosyncratic, the particular, the focus on the self as actor or observer or storyteller, as Rosenzweig and Thelen have astutely argued, define what is distinctive about this level of memory. Consequently memories of visits to the Fair may simultaneously replicate and resist the narratives created by custodians and institutions of the collective memory around the St. Louis Exposition.

I begin with oral histories of the Exposition because they focus attention on individual experience rather than broad collective memories developed largely by St. Louis memory institutions. The St. Louis Exposition is fortunate to have as source material a number of interviews with surviving visitors, interviewed and recorded in 1979 by the Missouri Historical Society. In addition, there were a number of reminiscences published, from time to time, in the *St. Louis Post-Dispatch*.[20] Unfortunately, this cohort of 1979 interviewees represents a special group who were either young children or in their early teens when they visited the Fair. Because of this age inflection, we might expect—and should not be surprised to find—almost no commentary on the major exhibits and silence as to the ideological lessons preached in the anthropology section. This was a silent reminder that the Fair was viewed differentially by different ages. Young people's experiences appear to have been centered on the Pike and are generally recounted in the context of who accompanied them and how they got to the Fair. There are occasional scenes that flash to mind, and the recollection of the night lighting, for example, or various rides and concessions. Recollections among this group also suggest that experience was gendered, with some itineraries clearly appealing more to men than women. Although the Filipino Igorots were mentioned frequently, as well as the story of their exotic diet and scanty clothing, none of the speakers contextualizes the story within the anthropological lessons promoted by the Fair designers. If they understood the racial and evolutionary context, they either neglected to mention it or had forgotten. Occasionally there are fascinating stories that have the ring of certain truth; for example, of how youngsters sneaked through the fence to avoid paying the entrance fee. And there is a long and instructive description of one girl's summer job on the Pike as a "booster" (a shill placed in a crowd listening to a barker who encourages customers to enter) for the Moorish Palace.[21]

By and large, these transcribed memories of the Fair have not suggested to historians any new ways of understanding the Exposition. Perhaps their transformation into amusing or mildly shocking stories seems to render them unusable. Even the representatives of the Missouri Historical Society who took the recordings voiced doubts about the authenticity of some of the recollections, worrying that they were embellished for dramatic effect. One man, they note, spoke without stopping for twenty minutes or so, as if he were "speaking from notes." It seemed to the interviewers that this made his account suspect and possibly a composite of his and others' experiences. In some cases, the two interviewers concluded that they had asked the wrong questions or even intimidated interviewees because of their apparent official status.[22]

How these memories might be useful in my exploration of experience must wait for later comment, while I explore other problems of the memory culture from the Fair. But obviously, what we can learn from such sources and how we understand memory depend upon understanding memory's construction and preservation. This is a process fraught with problems. In assessing oral accounts and other individual memory sources, for example, there is probably no practical way to untangle the effects of anticipating and remembering the Fair. The same documents, news stories, guidebooks, and photographs issued throughout 1904 did double service, preparing visitors to see the Exposition and describing what they did see afterward when they revisited them as souvenirs. Attention to the interconnections between anticipation and remembrance suggests how the various messages and interpretations of the Exposition were reinforced and how they might have modified individual experience.

There is also a problem of exaggerated scale. Children's memories undoubtedly envision a world where the hills are steeper, the distances greater, the crowds larger and more confusing, and the times more exciting than they might have been for an adult. As historians looking at recorded memories, we need to ask whether this may not also be true of collective memory in general, where the past itself is transformed by the years into something like the child's awe: more remarkable, inspiring, interesting, or perfect than it could have been—or even seemed at the time.

The transformation of experience into some form of recollection is a universal process, as is the influence of multiple sources upon the resulting meaning. This process is rarely visible or obvious to historians. Thus, we can only speculate about how this process works, because we so rarely see it happen. Fortunately, in Martha Clevenger's *Indescribably Grand*, a useful collection of letters and diaries regarding the St. Louis World's Fair, the transformation is clearly visible in two striking documents: the diary of visits to the Fair by Edward V. P. Schneiderhahn, and a more formal reflection by the same author entitled "World's Fair Memoir," written somewhat after the event. Schneiderhahn, the son of a German immigrant, was a St. Louis attorney in 1904 when he visited the fairgrounds on more than twenty occasions during the spring, summer, and fall. During or after each visit, he made a brief diary entry listing the exhibits he saw and including a few attenuated reactions. Although special occasions such as German Day and St. Louis Day were events that drew him to the grounds, he spent his time more or less equally divided between concessions on the Pike, the central exhibit halls, and attending parades and musical performances. The most immediately striking characteristic of these visits was their almost random, piecemeal, and arbitrary circuit. If Schneiderhahn was following any set itinerary, it was undoubtedly one of his design. Although many stops along the way may have been suggested by newspaper articles, guidebooks, and advice from others, or by photographs of intriguing displays he had seen, clearly this was an idiosyncratic selection. If the route of his visits seems almost serendipitous, this is because he did not follow the prescriptions and advice tendered by the designers and the exhibitors. In other words, his path reveals no larger ideological or cultural constructs save his own enjoyment of things German, and, on occasion, his temperament when he rails briefly against nudity on the Pike and in the Art Gallery, and about disrespectful French waiters. His itinerary is also characterized by the brevity of each visit. Taken together, his tour may well reveal a pattern of jumbled and shifting attention that was typical of many other patrons. With so many curiosities vying for attention, with halls crammed with eye-catching displays, this was probably inevitable.[23]

Nonetheless, when Schneiderhahn came to assess the meaning of the Fair in his "Memoir," his remembrance suddenly assumed a new, and orderly, shape. For this different literary format, he reorganized his experience and added historical perspective, including descriptions of sites he may or may not have witnessed (several are not mentioned in his diary), at the same time passing over whole sections of the Fair that had occupied much of his time. Indeed, the "Memoir" resembles nothing so much as a guidebook or officially provided news release. As is generally true of such publications and this genre of prescribed descriptions, his account moves from general, large-scale perspectives to narrower discussions of individual buildings and then, within them, to certain highlighted displays. In other words, he reordered and contextualized his experience within a familiar format. Accordingly, his account moves from the United States Government Building to Liberal Arts, to Manufactures, to Varied Industries, Transportation, and so on through the hierarchy of main pavilions. Within this geography of exhibits, Schneiderhahn rewrites many of his diary comments and adds several new judgments. He remained dubious about some of the art on display. His preference for things German remains clear, but he now adds several comments about the impressive progress of the Japanese. While many of his positive reactions to exhibits remain quite personal, he repeats the intended lesson of the Philippine exhibit: that the display "proved the high civilization already attained" under American tutelage. The account concludes with a note on the "beauty of the [nighttime electrical] illuminations," thus ending his story in dramatic fashion, with the fall of night, the implied passage of time, and the end of the Fair.

The most striking element of this memoir is the absence of any mention of the Pike, where Schneiderhahn spent a considerable portion of his time.

This omission is even more unusual because the memoir is a reflection on his diary, in other words, an interpretation of this daily accounting of his life. Probably intended for his own use, this revision of experience is clearly a document that revised the way he visited the Fair into a narrative that emphasizes the approved cultural, intellectual, and ideological purposes of the Exposition and that incorporates his visits and immediate reactions within this new framework and with dramatic effect.[24] We see, in other words, that this transformation places his individual experience within a generalized presentation: it restructured his memory to fit inside what was clearly intended to be the lasting impression of the Fair that coincides with the purposes established by its creators. Why did he omit the Pike and his enjoyment of its many concessions? We cannot know, although perhaps he was only reflecting the attitude of the management—which de-emphasized the Pike except privately, while depending utterly on the proceeds of its concessions. Even if entertainment drew him to the Fair, he does not admit this in his memoir. While this is clearly an idiosyncratic manipulation of experience and memory, it can be instructive in a larger sense. It may generally be true of oral history as well as memoir literature that it is susceptible to a similar dialectic process of encounter between individual experience and the cultural instruments of meaning present before, after, and, of course, everywhere on the fairgrounds. Or perhaps he was only anticipating the expectations and judgments of a possible audience for his memoir

Comparing Schneiderhahn's diary and memoir reveals one way that experience mutates into memory. I wish to turn briefly now to consider how the huge output of print and photographic material presented the Fair for prospective visitors and for those whom I will call "distance spectators," in other words, those who did not travel to St. Louis but became acutely aware of the Exposition through publicity and thus developed a knowledge (or secondhand memories) of it. By and large the most important forms of this print culture were found in newspaper and magazine articles and guidebooks. Print genres reinforced the messages and hierarchies promoted by the Fair management, although they sometimes rearranged the order, emphasizing one display or another. Often they passed judgment about the architecture, discussed the urban ambiance of St. Louis, or highlighted particular exhibits for local interests. Newspapers and magazines frequently reproduced interviews with key figures planning the event or, as I have shown, allowed some of the Louisiana Purchase Exposition Company men to contribute articles describing the highlights of the Fair or its anthropological exhibits. This literature is both vast and frequently repetitive. It is also filled with subtle variations that make it impossible to summarize succinctly. Nonetheless, certain consistencies emerge in this publicity that may well have influenced visitors and the memories that emerged both individually and collectively from the Fair.

An interesting and frank assumption about the importance of print media to recollection is the frequent mention in contemporary souvenir guidebook titles of the word *memories*.[25] At the same time, we need to be cautious about ascribing direct and lasting influence to mere exposure to such publications. It is possible and probable that distance spectators, like the actual visitors to the Fair, fit their reading and viewing into their own intellectual itineraries, rather than simply accepting the arguments of the books, articles, and souvenir photographs to which they were exposed.

Both oral history and memory are subject to the crosswinds and influences of many distractions and influences. Some of these blow intentionally. The memory institutions created by the leading citizens of St. Louis after the Fair had the distinct purpose of establishing the historic meaning of the Exposition and guiding local memories of it. Not surprisingly, David Francis was the key figure in founding the most important memory institution after the Fair. But Francis also published a two-volume history of it in 1913. In the immediacy of the event, this semiofficial history continued to promote the universal vision of the Fair and its ideological and cultural claims. The memory organization that Francis founded, however, increasingly came to stress the more parochial concerns of the city of St. Louis.

Francis's two-volume *Universal Exposition of 1904*, published by the Louisiana Purchase Exposition Company, doubled as an official history as well as an interesting justification of the ideology that articulated the displays throughout the grounds. In a curious, almost defensive, preface to the work, Francis wrote of the exhibit as a kind of Utopia. He elaborated this point by arguing that there had been no discrimination against women or choice of favorites among European nations. Segregation and special treatment were applied only "in the case of those people close to or in a state of barbarism." Hence, there was "no negro exhibit as such, just as there was no woman's building to suggest the separation of sex or assume the inferiority of one as compared with the other." Evidence suggests that Francis did, indeed, make some efforts to tone down overt segregation on the fairgrounds. But his concept of inclusion could also be a means to repress overt expressions of diversity. And the category of barbarism, however one might define it, was certainly a pretext to segregate. The largest achievement of the Fair, he wrote, was that it summarized "the best of man from the beginning of society." Considered together, the displays represented the pinnacle of human achievement. In other words, Francis subscribed to the explanation that placed American history at the goal end of human evolution.[26]

The copious photographic program of these official volumes reinforced the grandiose vision of St. Louis as the apotheosis of civilization. Beginning, as most guidebooks had done, with panoramic perspectives and scenes of the stately central exhibition palaces, the book also included a large number of individual portraits of those who worked in the corporate administration, a compendium of the St. Louis elite. Other popular visual topics included the abundant statuary on the grounds as well as shots of activities on the Pike. Other photographs included one circulated frequently elsewhere: Pygmies "reenacting" a decapitation. At the same time, there were few interior displays of goods, merchandise, or machinery and almost nothing to suggest the huge exhibit space given over to agricultural products and livestock. Thus, in concentrating on the ensemble of buildings and their embellishment, the photographic insertion reinforced the primary message intended by Francis: that the Fair represented a stage upon which the highest achievements of civilization and industry could be glimpsed.

In a gesture of greater importance than any memoir, Francis worked assiduously to preserve the company records of the Louisiana Purchase Exposition Company; and with this effort he also sought to maintain a vivid memory of the Fair that was accessible and engaging to citizens of the city. He hoped his endeavor would create a usable past for St. Louis, a sense of city history that could be invoked as an argument for investing in progress—in the sense that a better past meant a better future through continued enlightened leadership. At the same time, he and the groups and individuals that followed in his footsteps created a highly selective memory that eventually localized the significance of the event and radically altered its meaning.

After the Fair closed in late 1904, Francis and his fellow commissioners of prominent St. Louis bankers and businessmen planned to establish a Louisiana Purchase Historical Association for the preservation of documents and for the furtherance of Fair memories. Although this organization was not officially founded until 1916, it was, in fact, an extension of the Louisiana Purchase Exposition Company, which functioned until 1911. In 1910, the company formally approved the construction of a Jefferson memorial (on the former grounds) as a memory of the Fair and a monument to the "spirit of civic pride that made it possible." The building also served as a storage facility for the records of the company, papers and memorabilia associated with the Fair, and the core of the collection that the Louisiana Purchase Exposition Company later donated as part of a "historical museum and library." With $100,000 in surplus funds earned from the Exposition, the company set up an endowment to support the project. When the Louisiana Purchase Historical Association was formally chartered in 1916, it elected Francis as its first president.[27]

Before this, the first project of former Louisiana Purchase Exposition Company members was not the historical association, but plans for a St. Louis Public Museum that would house and exhibit a portion of the vast collections from the Fair. This was, in effect, a mini extension of the Exposition. A meeting of the Louisiana Purchase Exposition Company in November 1904, prior to the closing of the Fair, established an executive committee consisting of representatives from local libraries, the Missouri Historical Society, Washington University, the mayor's office, and local businessmen. Their task was to locate an exhibit hall on the fairgrounds to house the collection of exhibits that might be available after the Fair ended. Walter B. Douglas, president of the Missouri Historical Society, stated the obvious with his early warning that "the dissipation of the great collections costing millions and forming the greatest educational agency the world has seen will begin within a fortnight." Anthropologist William J. McGee was made managing director of the projected museum at this meeting, with an annual salary of $5,000.[28]

McGee was energetic in tracking down possible objects and artifact donors, but there were competitors from other museums interested in the leftover displays, and there were large costs involved in buying sizable collections from exhibitors. Nonetheless, the St. Louis Public Museum secured space in the temporary West Annex of the (permanent) Palace of Fine Arts on the fairgrounds and proceeded to mount those exhibits that could be secured. By opening day in July 1905, the museum had amassed enough to fill several exhibit rooms. Although initially crowds were large, particularly the first two days of July, attendance figures by December 1905 had fallen off severely.

The collection itself was eclectic, a sampler, in effect, of whatever the directors could acquire. There were scale models of the Temple and Tabernacle of Jerusalem (from the Jerusalem concession); "Klondyke" minerals, including a large gold-copper nugget; relics from the Robert Burns Cottage Association; a Japanese Temple; a model coal mine; a Philippine exhibit of sugar cane mills; Argentine leather; Chinese statuary; and so forth. The museum offered itself as a site for lectures and meetings, and several were held on anthropology, dental science, and other topics.[29] With flagging attendance a problem, the chief obstacle was the annex itself. Never intended to be permanent, it began to deteriorate dangerously. With revenues declining in 1906, McGee was forced to raise money to pay his own salary and then agreed to accept half his contracted income for several months. Even though the museum continued to exist in name for a few months longer, it was obliged in October to vacate its crumbling quarters and begin dispersing its collection. The Palace of Fine Arts building itself was not available, because it was slated to house the St. Louis Art Museum collection, and efforts to find another venue failed. If

McGee's purpose had been to continue educating the public about the evolution of the races and cultures through this exhibit, his efforts only resulted in a temporary jumble of tempting leftovers.

Because the charter of the Louisiana Purchase Exposition Company authorized it to establish "museums, libraries, art galleries, or the erection of public monuments," the executive committee eventually granted $100,000 in funds for the erection of a permanent Jefferson Monument building in Forest Park. The building was dedicated in 1913, and temporarily housed the archives of the company and what remained of the World's Fair collection.[30]

In his long-continuing work with the Historical Association, Francis actively sought to preserve memories of the Fair as a form of civic inspiration. For example, in 1921 when the Missouri Theatre showed a film of composite photographs taken at the Fair, Francis addressed the initial session with long quotes from his opening-day speech at the Fair itself, predicting a huge success. Francis rehearsed his vision of the Fair as a turning point in the history of St. Louis and the nearby West. For schoolchildren, who were too young to be among the fortunate witnesses to the grand triumph of their city, Francis persuaded the theatre to grant free admission.[31]

After the official founding of the Louisiana Purchase Historical Association in 1916, the executives worked closely with the Missouri Historical Society in St. Louis to find a permanent disposition for the archives of the company. By 1922 the two organizations had begun to work out a merger, with the Historical Association funding three partial salaries at the Society. The union took place in 1925; the collections and archives now had a permanent home and much improved curatorial supervision. This brief chronicle illustrates Michel-Rolph Trouillot's argument that the making of archives involves a number of operations that select, limit, rank, and shape what is saved. As he writes, "Power enters here both obviously and surreptitiously."[32]

It is worth noting at this point something about the development of memory and the changes in the meaning of the Fair that accompanied this institutional development. As time passed, the Fair became associated more exclusively with the history of St. Louis. Consequently, its symbolic importance as a great shining moment in city history gleamed brighter than ever. In 1911, in a speech at the annual meeting of the group that became the Historical Association, David Francis spoke nostalgically of those great, shining days. His words anticipate the distillation of a generational memory and the hint that, perhaps, the time of such enterprises in the city might be over. "The success of the enterprise seems to grow greater as it recedes," Francis remarked somewhat sadly, "and its magnitude has appeared to increase so that no other city has had the temerity to attempt to equal it and none can surpass it."[33]

While there were frequent press notations and commemorations of the Fair up to the beginning of World War II, it was the Sally Benson series of eight short stories, entitled "5135 Kensington," appearing in the *New Yorker Magazine* in 1941 and 1942, that significantly and permanently altered the collective memory of the Fair and, for that matter, the meaning of turn-of-the-century St. Louis itself for both the citizens of the city and all Americans. Benson (who was only six in 1904) published the entire collection in 1942, calling it *Meet Me in St. Louis* after the popular contemporary song. This became the working title for the movie that MGM commissioned that same year.

Benson's stories, which provided the basis for the film, revolved around two intertwined plots: the romantic escapades of the Smith family children and the possibility that the family would move to New York, where Mr. Smith has secured a lucrative new position. Although the events of the original story took place in 1910, her fictionalized version was set back in St. Louis in 1903 and 1904, so that she could use the coming of the Fair to advance the plot. By January, in the stories, the family has begun to visit the burgeoning fairgrounds. Romances among the daughters and neighborhood boys continue to develop. The argument about staying and leaving intensifies until, finally, the children convince their father to reject the New York offer and remain in St. Louis. If anything, it is the Fair that makes this decision plausible; it brings the world of culture to the city: "The miracle of the World's Fair in St. Louis, rising as it did out of the wilderness."[34] In a way, this is a paraphrase of the most powerful reminiscence about the Fair: you can have the modern world, right in your own community, "Right here in St. Louis," and still keep the family together—or at least so it seemed possible in 1904.

One important element in Benson's account is the extensive use the family makes of the fairgrounds, in particular the Pike, with its various entertainments, and the encounter with people imported for display. There are numerous jokes about ethnicity—the Chinese, the Filipinos ("the Igorots, the Moros, the Bagobos, and Fiji Islanders, had been brought there to be stared at and not to be educated"). So too there was ritual mention of ice cream cones. And there was one important footnote to prevailing popular culture when daughter Rose plays the piano and Tootie (the youngest child) dances the cakewalk, a popular dance step at the time.[35]

Benson's account froze the Fair in a simulacrum of the happy past, sketched in terms of an extended Victorian family, budding romances, and a moment when the world came to St. Louis and St. Louis sparkled with possibilities. It footnoted the racial and ethnic attitudes prevalent during the period, although it transformed them into humorous encounters, but took no notice of the elaborate scientific framework created by McGee and affirmed by the

Fair administration. While the novel described the extravagant central display buildings, the attention of the family and their friends was clearly the Pike and its popular entertainments. In fact, in Benson's stories the Fair had become a vast middle-class amusement park.

The film *Meet Me in St. Louis*, shot in Hollywood during the end of 1943 and the first half of 1944, completed this process. It opened in St. Louis in December of that year—a brilliant assemblage of Hollywood clichés perfectly suited to a time of high anxiety. Judy Garland starred as the daughter Esther, together with Margaret O'Brien as Tootie, her younger sister. Garland sings a number of songs in the film and, in one memorable dance, performs the widely popular cakewalk tune from 1904 "Under the Bamboo Tree." The film cut through the novel's thick description of family life to focus on the romances of the children: Esther, her sister, and her brother. Probably the most important change is the relatively diminished segment devoted to the Fair itself and the complete excision, except for the cakewalk, of any direct reference to the highly charged anthropology of the Fair, which was vividly present in Benson's memoir. While there are suggestions of diverse ethnicity, these are much more in keeping with the 1940s tendency to footnote only European ethnicity and religious variety in American identity.[36] Something else should be noted about the impact of this film. Opening in 1944, at the height of World War II, this context intensified the nostalgic content of the film. In 1944 boys weren't leaving for Princeton Junction (as in the book and film), but for the war. And the "Boy Next Door" about whom Judy Garland sings so plaintively probably wasn't next door at all but on a troop ship heading toward combat.[37]

The war is emotionally present in other ways that may or may not have been apparent to the audience. In one strange and extended scene, a Walpurgis Night rendering of Halloween, Tootie (played by Margaret O'Brien) and a gang of her friends determine to "kill" a neighbor whom they dislike by throwing flour on his face to render him a ghost. Tootie accomplishes her scary mission against the neighbor, Mr. Braukoff, who looks suspiciously as German as his name. This is a marked translation from the original story, where the neighbor has the unusual Scottish name Waughop. There are other odd elements to the film that suggest that the family, the Smiths, might in fact stand in for the French aristocratic heritage of the city, such as Rose, the eldest daughter, speaking French.[38]

In other respects, the film deeply inflected memory of the Fair because it added a universalized and generalized nostalgia for the past, evoking not just the city but, in general, an era of good feelings, joy, and innocence for a modern generation that knew only depression and war and, for the citizens of

St. Louis, the increasing recognition that the city had lost out to its competitors. The past looked especially bright as the present and future dimmed.

Frequently thereafter, at commemorations and remembrances and exhibits, the film was presented as if it recorded a valid part of the city's memory, only increasing, it would seem, the power of nostalgia. It might be added, finally, that this was Judy Garland's second film about an American world's fair. The first was *The Wizard of Oz*, a 1939 movie based on Frank L. Baum's imaginary trip to what might be construed as the magical (Emerald) city of the Chicago World's Fair of 1893. Perhaps the audience for these films did not specifically make the connection, but both were packed with an emotionally charged nostalgia for Victorian America and Hollywood's iconic period of idealized families. It is also fascinating to note that Irving Brecher, the screenwriter for *Meet Me in St. Louis* also worked on the script for *The Wizard of Oz*.[39]

In the end, perhaps the most significant lines of the film are spoken by Judy Garland. Marveling at the Fair, she exclaims: "I can't believe it. Right here where we live. Right here in St. Louis!" Perhaps it was these words that most appealed to the later generations who saw the film.[40]

From World War II until the present, the memory of the Exposition has continued to play a leading, if special, role in the civic activities of St. Louis. One of the most interesting efforts was begun in 1945 by the Board of Trustees of the Missouri Historical Society. To commemorate the 50th anniversary of the Fair, and the 150th of the Louisiana Purchase, the board mulled over mounting another world's fair (in 1953) until the impossibility of such an endeavor became clear. As the *St. Louis Globe-Democrat* reported in July 1946, the moment the proposal became known, "nostalgia of a kind unequaled in this city or state broke loose on front porches, at clubs and in conversations among older businessmen who found out convincingly just what a large-scale fair can mean to a locality—and to a nation."[41] But nostalgia could not substitute for dollars and civic dynamism. And there was also enough outright opposition to block the project for fear of misplacing precious resources and the potential environmental damage to Forest Park again.[42]

The spring of 1967, when the Montreal World's Fair opened, provided another citywide occasion to revive memories of St. Louis's grand days past. The women's committee that staffed the visitor's center of the Old Post Office planned a celebration on April 15–16 to "recreate in miniature" the Fair. They gathered souvenirs, planned exhibits and entertainments, and encouraged visitors to come in old-fashioned dress. They also planned a commemoration of the Olympics. Ice cream cones were on sale for a nickel each.[43]

While St. Louis newspapers continued to mark anniversaries of the Fair with special features and theaters showed *Meet Me in St. Louis* to remember the Fair, it became clear that the memory of the Louisiana Purchase Exposition had begun to serve distinct political purposes in the city and to address different communities differently. Nostalgia for past triumphs, symbolized by ice cream and iced tea and other "inventions" of the Exposition, served those most associated with the populations that mounted and attended the Fair or belonged to the elites who acted as officials. Although there were occasional references to ethnicity and horrified comic stories about Igorots, the modern city of St. Louis was increasingly inhabited by populations who were underrepresented as patrons of the Fair or who had been unwelcome there and who certainly would not appreciate the joke.

The 1979 commemoration of seventy-five years celebrating the Exposition was the occasion to mount an extensive and long-running exhibit of artifacts from the Fair. Just as important, the Missouri Historical Society began a systematic, if late in the day, recording of oral histories of visitors to the Fair who still survived. The Society also prepared a number of special exhibits based on its collections to run in conjunction with other commemorations of the Fair held in Forest Park from April 30 to May 14. An obvious question that lay behind these celebrations was whether or not the collective memory of the Exposition was truly collective and not a past belonging to a small group with a special stake in that memory and the culture and politics it represented. That problem suggests why the events of 1979 and exhibitions afterward began to address more broadly based entertainments designed to attract a much wider audience and to make the memory of the Fair relevant to those groups that had not been among the original patrons—and, consequently, to dilute the historic content of these celebrations.

The Seventy-fifth Anniversary Celebration of the Fair in 1979 revealed these civic uses of memory and the mixed purposes of commemoration. The fete included some elements specific to the Fair, loving reference to the past itself and the "good old days," as well as events and entertainments designed to attract a contemporary audience with no particular knowledge of the Exposition. In conjunction with the Department of Parks, Recreation and Forestry and the St. Louis Visitor's Center, the commemoration featured band concerts, symphony orchestra performances, parades, a sampling of foods, and a 1950s documentary film, *St. Louis—That Fabulous Summer,* by Charles Guggenheim. Although the planned craft demonstrations, canoe flotilla, and sailboat regatta had little to do with the events of 1904, they provided entertainments for an audience that was, by 1979, almost entirely without any direct experience of the Exposition. With artifacts and collectibles from the period on display in the Historical Museum and other sites, there were costumes, photographs, furniture, and other memorabilia from 1904. Other St. Louis institutions, such as the Arts and Education Council, the art museum (occupying the only standing building in Forest Park from the Fair), and the zoo, with its bird cage from the Fair, participated in the summer-long celebration. The event also included several days of discussion about contemporary environmentalism and space exploration, plus a golf tournament and a national tennis match for players over thirty-five—activities unrelated to memory but consistent with the notion that this was an occasion to celebrate the city of St. Louis through entertainment and edification.[44]

Although the original Fair had focused much of its attention on distinctions of ethnicity and race, these elements persisted but were transformed in the 1979 celebration into rituals of inclusion, not exclusion. Among the entertainments were plentiful representations of ethnicity and race, including Polish dancers, blues singers, ragtime performances, Ozark music, and the Folklorico Azteca. The most frequent appearances, however, were by German and Irish groups. And the highlight of the celebration on June 16 and 17 featured a "re-creation of the Pike area with a German Beer Garden and an Irish Village." Perhaps they reinforced the continuing sense of the centrality of these two immigrant groups to the history of St. Louis and their special memories of the Fair itself.[45]

Considering the events as a whole, there are several conclusions that can be drawn about the function of commemoration and also about the role of the Missouri Historical Society in preserving the memories of the Fair. Most striking about this event is what was chosen to display and celebrate and what was not. In fact, the selection of memorabilia was eclectic, and the entertainments were often period pieces merely symbolic of the past, such as the ragtime band (ragtime had essentially been excluded from the Fair) and the antique auto show. In some respects, however, the commemoration preserved the innovative technological spirit of 1904 by featuring space exploration and displays of lasers and computers, in other words, the future of scientific development and consumer items. The oversized emphasis upon entertainment also reflected, in fact, what the original Fair patrons understood about its importance. Most significant, the commemoration suggests what its planners and designers believed would appeal to the existing public memory of the Fair. In other respects, however, this celebration initiated a challenge and correction of the past by seeking to construct a new and more inclusive memory. In this sense, it contradicted the original racial and imperial aspirations of the 1904 Fair.

Another conclusion has to do with the role of the Historical Society. As holder of both the manuscripts and documents—the historical source material of the Fair—as well as memorabilia in the museum, the Society served two principal constituents and straddled two foci of interest: historians and the public, history and memory. This dual role is readily apparent in its principal publication, the *Gateway Heritage* magazine, founded in 1980, and in its new online *Voices*. Featuring articles by historians, graduate students, and experts in various areas of St. Louis history and memorabilia, *Gateway* straddles both professional history and memory constituents. The direction of the articles it has published on the Fair is decidedly toward inclusiveness of groups slighted or excluded from the fairgrounds or from full historical discussion. And it has also extended itself to translate the imperialism-racism narrative into a more popular format. Whether this has actually changed the collective memory of the Exposition is difficult to know, but the effort to challenge it is obvious. With a current circulation of just under twelve thousand, it is not clear how effective such efforts can be.[46]

If one of the larger purposes of collective memory has been to incite civic pride, then another has been to preserve objects collected from the Fair and maintain their potential market value. Just as scholarly collections conserve documents and records, so museums, historical societies, and private collectors preserve items from the Fair in what we might call the archives of memory objects. There is considerable overlap between the sorts of objects preserved in public and private ownership, with photographs, printed materials, diaries and letters, and other items appearing in all of these venues. It should never be forgotten that, while memorabilia have considerable value to historians, they often have real market value. In this respect, memory also has a commercial identity, and, in particular, objects from world's fairs have considerable value. Among all the possibilities, the St. Louis Fair ranks only behind the Chicago World's Fair, the New York World's Fair, and the Chicago Century of Progress of the 1930s in the economic value of its souvenirs.[47]

Perhaps responding to the heightened interest aroused by the 1979 commemoration, a new 1904 World's Fair Society was organized in 1982. As one of its founders wrote, the group was "interested in digging up the facts and setting the record straight" as well as, for those who actually attended, "sharing our memories." In 1986, the society began to publish a monthly newsletter, the *World's Fair Bulletin*. This reincarnation of the *World's Fair Bulletin* suggests the direction and uses of memory in contemporary St. Louis. Activities of the group included several sorts of public and private events. Frequent showings of *Meet Me in St. Louis,* discussions and lectures, small exhibits, displays of collectibles and lectures on collecting, picnics, and show-and-tell meetings made up their busy schedule. The newsletter also contains news about anything of interest concerning the Fair, such as unusual incidents or odd exhibits.[48] Also important were claims of "firsts" at the Fair, and thus unique to St. Louis, such as the ice cream cone and iced tea. The society also floated the suggestion that St. Louis might hold a world's fair in 2004 and perhaps invite the Olympics to the city in a reprise of their first appearance one hundred years earlier. Reprinting a portion of a *St. Louis Post-Dispatch* column, the organization probably meant to underscore what the newspaper writer suggested: "We could re-create the magic that was the 1904 World's Fair in 2004. We could re-create the pride that so many St. Louisans had in their city when the world's eyes were upon it."[49]

Interrupting this celebratory frame of mind, there was a sudden and out-of-character book notice that appeared in the October 1992 issue of the *World's Fair Bulletin*. Under the strange title "1904 Fair Receives Unfavorable Press in New Biography," an unsigned essay reviewed the recent biography of Ota Benga by Phillips Bradford and Harvey Blume. It related a brief outline of Benga's exploitation by W. J. McGee and the anthropologists who organized the various ethnic exhibits. Noting that McGee failed "to prove the complete supremacy of the Caucasian race," the author indicated (perhaps to justify his review of such a negative work) that the book had been mentioned twice in the *Post-Dispatch*. It also noted that Reverend Dr. Samuel Phillips Verner, who brought Benga to the Fair, traveled back to Africa with the Pygmy. While there, Benga re-created the anthropology exhibit by agreeing to sit in a wooden pen that he constructed to show his fellows. The review ended with a brief account of Benga's suicide.[50] This unusual article is fascinating principally for the sudden incursion of contemporary historiography into collective memory. It was as if the challenge had to be confronted. It is also noteworthy from the tone and tense of the piece that the author thinks of the Fair as a living, contemporary event, its reputation still subject to news stories and bad "publicity" in the press.

What is particularly instructive about this incident is its illustration of how and why collective memory forgets or excludes. That which was clearly important to the designers of the Fair was recalled only piecemeal and only insofar as it served modern purposes. Thus, the elaborate native villages and the elaboration of complex anthropological theories were largely missing from recollection in part because they could not serve the narrative that depicted the Fair a momentous event in the history of St. Louis that could serve modern purposes. In fact, reference to the "memory" of the Philippine Reservation is probably strongest in contemporary St. Louis in the name "Dogtown," referring to the present-day Oakland area of the city adjacent to Forest Park that purportedly supplied animals for Igorot feasts.[51]

By the hundred-year anniversary of the Fair, the 1904 World's Fair Society had continued to grow and maintain its various functions as memory and collector of objects of the Louisiana Purchase Exposition. Members set up a large tent at the citywide celebration in Forest Park and enthusiastically reported on the events of 1904, including the reenactment of the opening-day speech delivered by an actor playing David Francis. Even more had changed since the disappearance of any survivor who had experienced the Fair. It was also the case that in 2004 only one of the members of the executive board of the 1904 World's Fair Society actually lived within the city of St. Louis. All of the others resided in nearby suburbs or, in one case, in Illinois. If anything, this geographic dispersal symbolized the changes both in St. Louis itself and in the function of a collective memory that maintained visions of the 1904 city that scarcely existed one hundred years later.[52]

In considering the history of these remembrance organizations and their activities, we can understand some of the ways collective memory operates. Like the historiography of the Fair, the chronicle of collective memory also records shifts in attitude and interest over time as it responds to social and intellectual change. But memory has different functions than history, and, on those rare occasions where the two come together, we can sometimes detect uneasiness or perhaps even hostility and, at least, competition. Memory has functioned, particularly in St. Louis, as the assertion of a departed and much missed civic pride, a sense of potential that was disappointed and frustrated throughout the twentieth century. It also appears that collective memory has been, by and large, the possession of particular groups in St. Louis who feel that they have some stake in re-creating the vision of the city as a vibrant urban culture, filled with energetic German and Irish immigrants, and bustling with the business of westward expansion. For them, it is not the depressed, red-brick city left after the deindustrialization following World War II. St. Louis was a city that once was theirs, even if only a memory of that possession remains.

It is, therefore, all the more fascinating to explore what happened when history met memory as it did in the Missouri Historical Society display "Meet Me at the Fair: Memory, History, and the 1904 World's Fair," mounted in June 1996. Although the intent of this display was to engage the heavy burden of collective memory of the Fair and question its particular role in the historical consciousness of St. Louis, the planners knew they had profound competition, which they acknowledged with the title: a name that recalled Sally Benson's work, the Judy Garland movie made from it, and, particularly, the nostalgic song associated with both.

Prior to mounting the exhibit, the Historical Society undertook an extensive survey of St. Louis residents to test the presence and extent of particular memories in different populations. The same survey associates did a postexhibition evaluation that, in part, confirmed its original analysis. With a population in the city already almost 50 percent black in 1990, the survey underrepresented minorities in the persons it approached. Thus, the final results were skewed, even if still revealing.

The survey asked individuals to name the most important three events in St. Louis history. Most mentioned was the World's Fair, by almost 50 percent of respondents, and the second, the construction of the Gateway Arch in 1966. But the breakdown of these figures is what suggests several striking conclusions about audience and the nature of the Exposition memory. The first division is by gender. Women were far more likely to cite the Fair as paramount, while men put the Arch first. Even more revealing was the racial and ethnic breakdown of responses. Whites tended by 20 percent or so more than African Americans to mention the Fair first, and, among them, ethnic whites ("those whites with a particularly strong ethnic identification, such as German or Irish") mentioned the Fair by 20 percentage points more than that.[53] Much can be suggested about the strength and presence of memory in different populations of St. Louis, much that might suggest who went to the Fair and why. But, for the present, it is enough to note that memory of the Fair was divided in the city and, perhaps, even controversial.

Organized and designed by Katharine Corbett and Howard S. Miller of the Historical Society, the exhibit itself relied on a national panel of historical advisers including Rydell, Burton Benedict, David Thelen, Neil Harris, and David Glassberg, all discussed earlier in this book. The purpose was to encourage audiences to confront the nature of nostalgic memory as well as to understand how historical memory is constructed. As the proposal put it, visitors would come to see that memory and history, although often attentive to the same events and circumstances, "frequently rank them differently and extract different meanings."[54] The exhibit intended frankly to challenge visitors to "reflect on the many ways people fashion and refashion historical accounts to make sense of the past and help explain the present." In fact, this was not the first such effort by the Historical Society to confront memory with historical interpretation. In 1988 it had mounted a smaller exhibit of the fair relying heavily on historical works about the Progressive Era, and particularly followed Rydell's emphasis on the racial and ethnic dimensions of 1904.

Probably the most innovative element of the 1996 exhibit placed the sources and forms of history making and public-memory making within the same general context, hoping to encourage visitors to see them both as varieties of an imaginative reconstruction of events. The exhibit narrative followed the prevailing historical consensus about the Exposition, emphasizing the optimism of the Progressive age on display in 1904, the dream of unending technological progress, and the anthropological justification of America's then recent conquest of the Philippines. It re-created the self-image of America as the pinnacle of Anglo-Saxon civilization. It also underscored the practical purposes invested in the Fair by the city's forward-looking elite, who hoped to use the occasion to assert their reform style of government and local industry, to lead St. Louis into more progressive and prosperous times. In a fascinating insight into the Exposition, the authors recognized that these "ideological and political themes of the 1904 World's Fair were probably lost on most visitors, at least on a conscious level." They speculated that the appeal was, instead, the grandeur, innovation, and sheer magnitude of the Exposition.[55]

This determination encouraged the authors of the commemoration to pay particular attention to the way memories of the Fair were constructed and then functioned throughout the century since 1904. Although this attention to memory suggests that it might have opened the door to an examination of experience, the designers were generally more interested in understanding the social function of those memories: their role in creating a collective version of St. Louis history that inflected family histories, frequent commemorations, documentary films, public school curricula, and a growing collectors and souvenir marketplace. Corbett was also struggling against the stereotypes of official memories that still persisted. As she put it, "In a city with a well-defined business elite, the official memory continually reminded the populace of who had orchestrated St. Louis's finest hour."

Perhaps the most interesting and inventive segment of the exhibit addressed this bias with the impersonation of fifteen characters that attended or helped to mount the Fair. From all walks of life, both from visitors, but mainly from participants in the Fair (including Ota Benga, Scott Joplin, Jessie Tarbox Beals, and, of course, David Francis), these characters were chosen to suggest the variety of peoples and experiences gathered together on the grounds. The exhibits were also intended to reproduce something "reminiscent of the Fair itself: non-sequential, fragmentary, visually and artifactually dense."[56]

Finally, the visitor was meant to confront a number of displays that suggested the ways collective memory was developed and then exploited: through documentaries, official uses of history, and the role of collectibles and souvenirs. This was all extraordinarily ambitious. But what, in fact, in the end, did the 1996 visitor understand? In an article in the *Exhibitionist* in the fall of 1999, Corbett discussed the purposes of the exhibition and frankly evaluated its successes and failures. Based on a survey of visitors and their reactions to the exhibit, she found mixed results. A number of visitors resisted the multiple perspectives on the Fair, particularly those emphasizing race and class differences. Others apparently did not understand the problem that historians posed about encountering differing and opposing voices and interpretations. Representatives from the Philippines were one group angered by proposals to show the popular photographs of dog eating by Igorots. There followed a "long, painful negotiation that resulted in compromise."[57] And representatives of the 1904 World's Fair Society were unhappy with the (critical) historical perspectives revealed by the themes of the exhibit. Finally, Corbett was less than positive about the effect of the exhibit in encouraging visitors to discuss and understand the basic premise: the conflicted and complex nature of history making and memory making in all of their guises. Whether this result reflected the disappointed expectations of the visitors, certain design flaws, or lack of preparation for what was actually the purpose of the exhibit, she concluded that "in the case of *Meet Me at the Fair*, meaning-making was best provoked in the section where we encouraged visitors to consider how treasured family objects influence family narratives." Indeed, those with personal connections to the world's fair came away understanding better "why they cared about the fair as an historical event."[58]

In another sober evaluation, she and fellow author Howard Miller concluded that the exhibit asked visitors "to take the fun out of the fair, to trade enjoyment for analysis, heritage for history." This they resolutely refused to do. Patrons entered through exits, and left through the entrance; they ignored some artifacts and proved selective in what they wanted to see. Most visitors, she concluded, "left as they had come, chatting about collectibles that reminded them of the glorious summer when St. Louis was the center of the world."[59] This obstinacy affirms the well-known tendency of patrons to impose their own agendas on museum exhibits. This willful behavior may also offer insights into how visitors interacted with the 1904 Exposition itself.[60]

The evaluative survey upon which Corbett based her conclusions is also fascinating for what else it reveals about the audience. Counting the number of viewers for videos suggests that the two most watched films dealt with the Fair in movies and as a function of family memory. The least watched explored the discussion of the fair in history and "Selling the Fair." The most interesting conclusion came in the measure of "respondents' awareness of exhibition themes." This section collated responses to the six principal ideas stressed in the exhibit, in other words, those themes the planners and designers hoped the audience would understand. The survey found that about half of interviewees understood the notion of history as that which people choose to remember about an event. And a similarly high percentage understood that one's position in society could influence experience of the Fair. This suggests that most visitors grasped something, although not necessarily all, of the perspective being promoted.⁶¹ But the failure of many visitors to understand or respond to the general point of the exhibit (or at least articulate this understanding in expected terms), despite the concise and overdetermined didactic environment, is a lesson for our speculations about experience in the much more heterogeneous and diffuse environment of the 1904 Fair.

While this effort to influence and revise collective memory had its successful moments, it remains clear from a consideration of other events, that nostalgia and dreaming of a better yesterday still strongly tint the meaning of the Fair. The Judy Garland movie continues to evoke such memories, and, in 2004, the St. Louis County Library reprinted Sally Benson's novel on which it was based as part of the hundred-year commemoration. For those who wish to see the Fair as the "historic" part of a present-day amusement park, there is the Six Flags Over St. Louis theme park located in nearby Eureka. And the official travel Web site of the city advertises a "1904 World's Fair Tour in St. Louis" that touts the same stories of ice cream cones, the huge size of the park, collectibles, and an itinerary to the few remaining structures from the Fair. It would seem, then, that collective memory of the Fair continues to be energized and perpetuated by the very ideas that historians fault as nostalgia and fantasy.⁶²

This long excursion into collective memory and its organization and organizations should indicate that, although individual memory participates in it, this should not exhaust our survey. Personal memory is far more elusive, even if it is sometimes shaped and reshaped by collective recall. It is also difficult, if not impossible, to quantify, and thus can be recounted only individually. Nonetheless, there are certain generalizations that one can make about individual memories of the Fair, and they and (surprisingly) some elements of collective memory can be very useful in attempting to reconstruct the original experience of visitors in 1904.

If there is one constant in all versions of memory, it is the individual's need for a usable past. How he or she is personally situated in relation to collective memory is what ultimately matters. Understanding a similar relationship between subject and object, between the individual viewer and the viewed can help enlighten the meaning of experience on the fairgrounds in 1904. The idiosyncrasies of individual memory may therefore reveal something fundamental about the nature of experience.

Although the project of recording individual memories of the Fair has been uneven and historical interest in such reminiscences has grown only recently—unfortunately, after most of the visitors have passed away—there still exists considerable useful documentation. None of this material lends itself easily to generalization and thus can only be used impressionistically.

Such a qualification is certainly applicable to all literary sources. But with millions of different possible memories among the millions of visitors, those that we do possess must always be considered singular and quite possibly uncharacteristic. We also need to consider this information in light of theoretical hesitations about the validity and usefulness of oral history and individual recollection raised earlier in this chapter. Nonetheless, we can still speculate from the material in existence and attempt to read it alongside what else we know of the Fair, partly to restore the recorded voices of the audience and learn something of the idiosyncrasies of individual visits. With caution, then, certain types of general statements can be ventured, even if they cannot be numerically authenticated.

First we need to determine what does not appear in diaries or in oral histories: the significant missing information and lack of recall about some of the most important messages promoted by the Fair administration. There is little, if any, evidence that visitors remembered their experiences in terms of the hierarchies of knowledge and display that were promoted and visible everywhere on the fairgrounds. Nor did memories or diaries generally recount experience in terms of the suggested guidebook itineraries (which generally followed this hierarchical plan), even if some visitors might actually have followed some sort of prescribed plan of visitations. Their memories have lost structure and generally appear to be disorganized and impressionistic. Therefore, if the Fair constituted an encyclopedia of knowledge and if memory is a test of the learning that occurred on the grounds, then that learning emerged in jumbled, particular, and partial fashion.

This is partly what Martha R. Clevenger concludes in her fine discussion of diaries and letters from the Missouri Historical Society's 1904 World's Fair collection. She finds little discussion of ethnology in these sources and speculates that visitors probably found nothing surprising or shocking in the living tableaux of human cultures.[63] Quite possibly visitors did not question the depiction of racial difference and inferiority because they fit anthropological nomenclature into existing prejudices. But the anthropological orientation of the Exposition had many more complex ideas about ethnicity and evolution that were not reflected in these documents. In fact, the oral histories and diaries do contain important traces of this aspect of the Fair, but generally as amused or shocked reactions to the exotic and erotic aspects of these displays rather than the complex notions that McGee and his fellow anthropologists were attempting to explain.

The most obvious references to anthropology in oral histories are the tenacious stories about the Igorots at the Philippine Reservation, which still, it seems, serve in narratives to shock and delight. For example, Ernest Link, interviewed for the 1979 remembrance project associated with the seventy-fifth anniversary of the Fair, had a vivid recollection that he had missed seeing Filipinos eat dogs: "Now they talk a great deal about these Igorrotes that they had and their dog diet," he explained. "I never saw them eat, and I never saw them prepare a dog, but some people say that they did. We'll just have to take their word for it."[64] This negative iteration of an almost universal story was sometimes embellished with titillation about the native dress of the Filipinos. As Louis Venverloh explained, he was not allowed to see the Igorots because of their skimpy loincloths, although his mother and father went. But he did recount a chance meeting with an Igorot woman who "was cooking a pot of food, a great pot of food" that she admitted was "rice and dog."[65] Among the stories about the Fair, this was one of those most frequently mentioned in oral histories and a staple of the memory of the Fair, often recounted as if it were a shocking revelation. What we do not know from this passage is whether the encounter ever occurred or whether it was Venverloh's imaginary effort to put himself at the center of the (forbidden) Igorot controversy.

What else can we learn from individual recollections of the Fair? Many of the oral histories and diaries record individual reactions to the concessions along the Pike. These are enormously varied and run from the prudish rejection of exotic dancing to memories of the encounter with the "Hereafter," a popular guided tour through Heaven and Hell. William Shelp, who was twelve in 1904 and a native of St. Louis, illustrates a double-vision variety of recall: his own experiences, including what he probably heard somewhere, which he then incorporated into his own story. His interviewers reacted strongly against this form of recall, wondering "how much of what he had told us was his own experience or that of his parents." His detailed description of an Eskimo man using a whip to bounce a coin off a block of wood into the air was an oddity he might well have witnessed. But his tale of scandalous love affairs that were revealed when the Ferris wheel broke down, trapping its passengers for hours in its swinging gondolas, is probably something he only heard about or read in the newspaper. Yet this sort of compound view reveals the inevitable ingredients of memory: individual experience, transformed by collective memory, and embroidered with interesting gossip and tall tales.[66]

One fascinating and oft repeated sort of information relayed by almost all oral histories was how the interviewees went to the Fair and who accompanied them. This disclosure can be very important (although it is generally overlooked), because it suggests the social setting for visitors as they encountered the Exposition. A number of respondents remembered that they attended in family groups, many of whom had traveled to St. Louis to stay with relatives. Emma Helena Kuhn Mohr, for example, was eighteen when she journeyed from Trenton, Illinois, to St. Louis to spend two weeks with her sister and family who had moved there. As I will discuss later, this form of family reunion played a considerable part in the meaning of the experience.

Considering both collective and individual memory accounts, we are compelled to conclude that the Fair has given rise to forms of meaning that rarely figure prominently in what historians have written in their evaluations of the event. The priorities and experiences suggested by these memory accounts do not reflect or repeat the lessons embedded, at every turn, in the planning and execution of the Exposition. Instead, memory appears to have its own agendas. It seems to possess a will and a history independent of the best-contrived plans to convey particular meanings on the fairgrounds. Despite the documentary evidence preserved in the archival record describing the intentions of the Louisiana Purchase Exposition Company, it appears that a great many patrons did not learn the lesson of the Fair. Perhaps this is so because, as Siegfried Kracauer notes, "An individual retains memories because they are personally significant."[67] And perhaps that is the key: what mattered was an individual matter.

Should we then ignore memory, reject it as a misunderstanding, perhaps fault or dismiss it as trivial and too hopelessly idiosyncratic to be useful? Or can we find some way to read through it, as we need to read through the historical record itself, to reassess the experience of the audience in St. Louis? Can we imagine a different sort of narrative that takes account of this experience?

Memory now and immediately after the Fair delights in originality: for example, the legend of the first ice cream cone and iced tea; the biggest world's fair in history; the heyday of St. Louis; the first Olympics held in the United States. It is etched in local pride. It overflows with nostalgia, much of that created by the Judy Garland film and the larger Hollywood project of mythologizing the past. Memory also helps to sustain the cash value for collector's items from the Fair, the souvenirs and collectibles that have a distinct and increasing monetary worth because of their age and authenticity.[68] It often varies with ethnic or racial or economic group. But it also tells us a great deal more than that. We must learn to listen, because it suggests what is emotionally and politically important about the past. In so doing, it hints at what may well have mattered most at the time.

CHAPTER FOUR

MAKING IMAGES

Photography viewed and consumed in its various formats and varieties is a crucial ingredient in memory; indeed, it can substitute for experience itself. It can alter or become a visible impression. It is also a primary source for historians requiring interpretation and attention to function and context. Photographs and stereographs taken of the Louisiana Purchase Exposition provided vicarious experience for those who could not attend, just as much as newspaper and journal articles and novels and guidebooks did. Published images shaped expectations and affected experience itself, refiguring what had been seen, depicting what should have been seen, or offering an ideal perspective that only the photographer could record. Photographs embellished memory by substituting tangible and reproducible pictures for fleeting mental images. For the historian they are a complex, often underutilized source, filled with information about the Fair: how visitors might have envisioned it and details about the cultural interactions that made up the basis of the experience.

Photographs relating to the Exposition differed extraordinarily, ranging from the grand shots and long perspectives taken by the official photographers and distributed by the Fair management, to unique individual collections taken and also sold to the administration by professional photographers such as Jessie Tarbox Beals. There were the hugely popular stereographic sets sold by mail-order companies like Sears, or traveling salesmen, or at photographic stores; amateur and professional photographs converted into postcards; and individual shots taken by "camera fiends" and intended for family and friends. Given their different purposes, levels of professionalism, and formats, these images lend themselves to very different impressions of the Fair. They invariably suggest that an enormous investment of meaning can be returned by the simplest image. The Fair management was well aware of this power of the visual and so they distributed, in addition to their prose descriptions of the Exhibition, thousands of photographs, cuts, and plates. They also prepared stereopticon slides for "scores of lecturers" who performed in such venues as Madison Square Garden and Atlantic City to advertise the Fair.[1]

Sometimes the act of "capturing" an image spilled over with significance that went far beyond the resulting photograph. For example, Sam Hyde, a visitor to the Fair, recorded his attempt to take the picture of a Native American, an "old savage in all his glory of bead and feather." Hyde "knew" the risk in capturing this image with his portable Kodak, but he persisted: "'There's my game,'" said I, "'I'll shoot at him if I loose [sic] my scapp [sic] for it.'" The photographer ran ahead, positioned himself in front of his subject, and snapped the shutter. "As he stopped in his tracks with a savage grunt I shot across the lawn. He stood there for a moment glareing [sic] at me and uttering grunts like a hog." The photographer escaped, only to find, when the shot was developed, that the Native American had removed his headdress.[2] Hyde's attempt at humor played upon the legend that Native Americans resented photographers and only posed for a fee. But he also portrayed the adventure as a hunt for wild game and, beyond that, a reference recollecting the long history of Native American–European warfare. One much noted function of photography was, in fact, to record such adventures in imagination made possible by the extraordinary diversity of peoples brought to St. Louis and by the experiences made possible on its premises, with the camera as a weapon of conquest.

Before turning to the different visual aesthetic made possible by stereographic reproduction, there are some broad divisions of perspective that I wish to emphasize. These are particularly characteristic of the two-dimensional images available at the Fair and reproduced in books, newspapers, journals, souvenir books, and other media that published visual material. While there are a great many examples of shots of buildings, machinery, gardens, and statuary, images focused on people most interest me here. These may be divided into categories relating to the perspective of the photographer and his or her attitude toward the individual people and crowds that made up the human story of the Louisiana Purchase Exposition. This angle of vision captures a point of view that becomes the perspective of the consumer of the photograph: the eyes, as it were, of memory.

Beyond those stately photographs of distant glimpses of buildings, monuments, statuary, and park settings preferred by official boosters of the Fair, where people are either absent or obscured, most photographs of the Fair fall into three classes, depending upon how they portray individuals. The first depicts single persons or groups observing something at the Fair or walking or sitting, but, generally, oblivious to the camera. These shots tend to present people as an audience rather than subjects and focus on the buildings, large perspectives, or quaint settings; in some cases, they pan the large crowds assembled for parades or performances of one sort or another. In this genre, individuality is incidental. Depicting the audience interacting with the environments of the Fair is their function. In fact, in many cases, the camera itself becomes a fellow member of that audience.

The second category is far more intimate and focuses almost exclusively on people or individuals within a specialized environment. This genre of photography, often known as a "portrait-type" could be shot either in a studio, as Jessie Tarbox Beals often did when she photographed Native Americans, or in posed pictures taken on the Philippine Reservation illustrating eating, warfare, and other elements of daily life. Sometimes these appear to be candid shots, but more often they are close-ups of individuals or small groups looking straight into the camera, a highly desirable pose because of slow film speeds. A second possibility included family photographs or shots of friends visiting the Fair. The purpose of these may have been to exoticize the circumstances. In many cases, these are posed, emphasizing odd juxtapositions and intentionally jarring combinations of people and objects.

The final sort is similar, but rarer, and records the intimate interactions of audience and subject, providing visual testimony of how people of different cultures might have interacted or, perhaps better, how audiences exploited the various displays of cultural strangers. These, in particular, suggest meanings that go far beyond our normal sense of experience on the fairgrounds and will require a separate chapter to explore them.

While there are many approaches to interpreting photography and its potential revelations and constructions, the works of Susan Sontag seem particularly appropriate for understanding the prevailing practices of the period following the turn of the nineteenth century. In her collected essays on photography, she spotlights the language associated with early camera use: the metaphors of the hunt, such as shooting, capture, and imaginary risk taking, exactly as Sam Hyde described his adventurous photographic encounter with the Native American chief on the fairgrounds. Beyond the hunt and the "photo safari" that is associated with tourism, the image, she writes, gives the individual an imaginary possession of a "past that is unreal" and records spaces in which individuals "are insecure."[3] As she notes, taking such a photograph transforms the insubstantial into something real. Snapping an image can be imagined as a way of taming reality, particularly in America, where a vast geographic reality and cultural diversity loom grand and disorganized. Through photography, she concludes, an object or a glimpse becomes part of reality and a substitute for that which then passes away. Indeed, the image can replace or even invent an experience and a memory.[4]

This apprehending (and inventive) function of the photograph is something that Roland Barthes also stresses in his very useful essays on photography. The type of consciousness a photographic image provokes, he writes, is not the sense of actually being present, but rather *"having-been-there."* This affirmation of circumstance, however, does not entirely define the significance of an image, for each example has several potential meanings which can be fixed only through linguistic intervention—in a caption or other form of explanation. To this I would add the context of other images and the order in which they are placed or perceived. Sometimes, as Sontag argues, verbal contexts can override the apparent visual meanings.[5] Thus, it is appropriate to establish the setting of photographs, their placement, use, and reproduction: where they are located, how their appearance may generate one meaning or another, and how the visual image interacts with any surrounding textual material. So, for instance, in his essay on photographic representation, John Tagg discusses the ways in which photography is incorporated into and functions within disciplinary institutions like law enforcement, for example. Or as Siegfried Kracauer suggests, photography is also bound up and constrained in time, just as fashion and style.[6]

Perhaps it is true, as Peter Burke asserts, that historians generally ignore images or, when they do discuss them, "tend to treat them as mere illustrations, reproducing them in their books without comment." Rather than using an image to generate new questions, he argues, they seek to confirm what is supposed or known from other sources.[7] As I will argue, not only has this often been the case in terms of histories of the Louisiana Purchase Exposition, but it is also reflected in the singular attention paid to two-dimensional photography, as opposed to stereographs. This focus has excluded a popular and quite distinct fashion of visual documentation of the Fair. Ignoring three-dimensional images excludes a major category of source material from the photographic and historical record. This omission has an obvious practical explanation. The visual effects of stereography are impossible to reproduce on a flat printed page, and the images, by their nature, occupy double the space of a two-dimensional shot, which both limits the size of the depiction and distracts attention from the subject it is intended to illustrate. But this is only an explanation, not an excuse.

The format of commercial photographic books depicting the St. Louis World's Fair followed an order that, if not prescribed, was certainly a tradition by 1904. The Columbian Exposition in Chicago in 1893 gave rise to a huge photographic record, and its publications generally moved within an established aesthetic sensibility. The Louisiana Purchase Exposition hired official photographers to record the building of the Fair and commemorate its progress. Stressing the huge structures and fine perspectives, gardens, and sculpture, these shots celebrated the vast size and multiplicity of exhibits. While there was some attention to crowds and an occasional focus on single persons, most examples emphasized architecture above individuals.[8] The standard guidebook program of presenting images tended to follow this same architectural priority. Initially locating the visitor in the garden setting of the splendid alabaster palaces of Industry, Mechanics, Transportation, and Liberal Arts, these two book genres moved the viewer down rungs of importance, picturing the visually most impressive structures and then on to other buildings: to the displays of France, Germany, Britain and other nations; to the state exhibits; and from there to areas of special interest such as the Pike and its concessions, the anthropological exhibits, and the military parades and other special events. The ending was usually a collection of odd views or exotic individuals. The White City Art Co., in its souvenir book *Snap Shots of the St. Louis Exposition, 1904,* for example, followed this visual itinerary, reproducing, as it were, the ideological values of the administration, which also prioritized the grandiose and monumental. The volume even relayed the theme of the Fair, declaring that it was "essentially an exhibition of processes." But like many of the photo books of the Fair, it also devoted considerable time depicting curiosities, in this case, photographs of Native Americans and other peoples brought to the Fair for show. Like other books of this sort, it located pictures of erotic interest in the midst of the exotic environments, in this case showing two bare-breasted Fiji women in a native village.[9]

One very noticeable attribute of photo books and guidebooks was the rare appearance of large crowds. In part this was inevitable, because many of these books were produced before the Fair opened to meet publication dates, or they were shot on Sundays, when the Fair was closed to the public. In some instances, even, photographs had to be enhanced with etched-in drawings of individuals strolling or sitting on outdoor benches. But beyond the demands of schedule, there were aesthetic reasons for this vacant appearance. As much as the prose descriptions of the Fair stressed perspective, grandeur, and unity of style, these visual accompaniments reinforced the notion of architectural uniformity and social order. Shots of milling crowds might disrupt this impression. It is also likely that the absence of crowds was intended to

FIGURE 1. The Palace of Electricity and the Palace of Varied Industries with diminutive figures. The strategy of photographers sometimes placed tiny figures into panoramic shots to emphasize the colossal size of buildings. (Official Photography Company; Missouri Historical Society, Photographs and Prints Collections.)

be a statement about the leisurely, contemplative, educational purposes of the central display sector of the fairgrounds. It is also possible that fewer visitors patronized these areas of the fairgrounds or were dispersed because of the greater distances between buildings. When the camera turned to depict the Pike, crowds and confusion returned. This aesthetic division reproduced and reinforced the separation of high and popular culture intentionally visible in the geography of the Fair. At the same time, it may have revealed the greater popularity of the Pike.

Collections by independent photographers also followed this visual itinerary. C. S. Jackson's collection of "famous" photographs from the Exposition also commenced with the grandiose and then devolved to other portions of the Fair. The first photographs included pictures of the leading figures of the administration, such as Francis. The collection then moved to some of the abundant statuary and monumental perspectives that photographers (and the Fair management) favored. The visual tour then stopped in front of international exhibition pavilions and state buildings and finally at the Pike. Once again, the collection put a number of shots of ethnic types from the various concessions and living tableaux of people toward the end. Commentary about the shots of individuals included discussions of Native

Americans and the usual allusion to Igorot dog eating. Next he pictured animals at the Exhibition zoo. The collection ended, however, on a high note after so much diversity of peoples and animals. The photographer concluded with a number of photographs of the scale-model reconstruction of the city of Jerusalem as if to sum up in a spiritual mode.[10]

Jackson's commonly repeated emphasis in picturing officials and official buildings at the beginning of his book was certainly a cliché, but perhaps it was also an effort to convince the purchaser and reader of the book of some sort of official sanction for the series. He was not alone in this practice. Several titles of photographic books and guidebooks included the word "official" whether or not they were actually approved by the Fair administration.

Like the guidebooks, which were often lavishly illustrated, photo books of the Fair described the visual program as equivalent to a trip around the world. They proposed to lead the viewer into a vicarious experience of the Fair. *Memories of the Louisiana Purchase Exposition* repeated the common theme of this genre. After an exploration through images of the popular spots along the Pike, the authors wrote, "you can feel that you have had quite an experience in foreign travel and know considerable about the customs of foreigners in their own countries." Another standard shot in this collection was of the audience. Like other photo books, *Memories* demonstrated two camera perspectives. The first depicted the audience watching, strolling, and engaging the Fair, often with their backs to the camera or seen from the side. Here, the spectators are shown gazing at scenes and people. The second broad category featured those observed, those encountered, many of whom faced the photographer and, hence, the purchaser or reader of the book. Faces among the vast crowds were unimportant and individuals were usually not pictured unless they were representative of something: an official, a person viewed as symbolic of a culture, or perhaps an entertainer or a celebrity.[11]

The Universal Exposition: Beautifully Illustrated reproduced the common declension from the grandiose to the specific that characterized many of these productions. It also follows one of the idiosyncratic visual deviations that several other photo souvenir books took. Because there was extraordinary diversity at the Fair, of exhibits, concessions, and peoples, some publications inevitably focused on subjects entirely ignored or underrepresented by others. *The Universal Exposition,* for example, reproduced photographs of the reenactment of the Boer War, a popular concession near the Pike.[12]

The photo album authored by Walter B. Stevens, secretary of the Louisiana Purchase Exposition Company, and William H. Rau, "official photographer," tried to capitalize on the impression of endorsement by the Exposition. The strategy of this volume was juxtaposition, something frequently found elsewhere in contemporary photojournalism. Many books like it reproduced figures of clothed members of the audience posed next to nearly nude inhabitants of the anthropological exhibits. Or they contrasted skin colors or remarkable variations in height and weight. Stevens and Rau used juxtaposition in their presentation to signify the remarkable comprehensiveness of types, cultures, and accomplishments of people at the Fair. Its all-encompassing unity brought together disparate human types and talents. "Finally," they wrote, "among all these strange peoples, civilized, semi-civilized and barbarous, assemble the thinkers of all nations to form a 'World's University,' 'to discuss and set forth unification and mutual relations of the sciences, to harmonize the specialized studies.'"[13] Matching this summary with visual examples, this unification through diversity ran from human cultures to the "encyclopedia" of contemporary thinkers assembled at the intellectual congresses organized by Munsterberg.

The authors articulated this program of diversity and juxtaposition in the order of their photographic choices. Beginning, as most books did, with the grandiose white display palaces, the book then alternated between showing spectators and buildings and vistas and statues, then exotic peoples, then more buildings. The longest section was devoted to pictures taken on the Philippine Reservation. But the larger strategy of the book was clearly to emphasize the variety of experience possible within one venue, something that could be envisioned, if not experienced, by this exercise in photo journalism.

Another souvenir picture book containing shots by Rau and his staff, *The Greatest of Expositions Completely Illustrated*, followed a similar zigzag itinerary. Its initial sections were devoted to the grand perspectives and palaces. But Rau seemed particularly interested in photographing the many American historical artifacts on display on the grounds: Abraham Lincoln's railway car, Teddy Roosevelt's "Prairie Cabin," the Liberty Bell, and "Battle Abby," an exhibit of famous American battle scenes. While showing some examples of foreign and state pavilions, the bulk of the book was devoted to scenes of the Pike and, finally, to the various ethnic and racial types so vividly present there and in the Philippine Reservation. A final shot in the book pictured African Pygmies meant to represent the lowest form of human being. This visual summary reinforced two notions: the comprehensiveness of the ethnic sampler available at the Fair as well as the hierarchies of "civilization" impressed upon foreign cultures by an imperial ideology.[14]

Three photographers, in particular, created a body of work at St. Louis that went far beyond these stereotypical portrayals of buildings, perspectives, and people. Their work was often reprinted in magazine articles about the Exposition or purchased and reproduced by the Fair administration itself. The

FIGURE 2. The art of taking pictures and the physical prominence of the photographer added to Beals's celebrity. (Jessie Tarbox Beals Collection, Schlesinger Library, Radcliffe Institute, Harvard University.)

first of these, Jessie Tarbox Beals, was one of the first nationally known female photographers, and her remarkable portraits and pictures suggest a different function for photography at the Fair. Beals had became a sort of celebrity journalist, sometimes as widely known for how she took a picture (from a hot air balloon, for example) and the extraordinary lengths to which she resorted to find the best angle, as much as she was famous for the finished view. Hired as a photographic journalist by several national newspapers, she gained permission to photograph widely on the grounds. The administration bought a number of her photographs, and the *World's Fair Bulletin* reproduced a good many of them in its later issues. While Beals took a number of photographs of dignitaries, she also assembled groups of shots into narratives such as "The

Children of the Fair" or "Strange Wedding Ceremonies."[15] In the next chapter, we will return to explore one of her most remarkable images.

The Gerhard sisters, Emme and Mamie, local St. Louis photographers, specialized, as did Beals, in close-ups and portraits. The photographs that they produced are especially focused on depicting the varieties of ethnic and racial types who peopled the ethnographic exhibits. They photographed and sold sets of images concentrating on exotic individuals and native peoples. The Gerhard sisters, however, often photographed these individuals in studio settings, allowing and apparently encouraging their subjects to look directly and boldly into the camera. This gaze back at the audience represented a distinct difference from other photographers, who tended to shoot such subjects from a greater distance and from the position of a hidden or unseen observer, to allow the viewer to overlook and inspect—rather than, in effect, confront—the eyes of the subject. Their removal of individuals from the cultural symbolism of display villages and reservations so carefully constructed by McGee and his associates, substituted a more general human-interest orientation for abstract anthropological concerns. Their single portraits of Native Americans thus contrasted with the strategy of other photographers who showed Filipinos, for example, in their compound, posed and holding spears, dancing, or roasting a dog.[16]

A good example of the latter was in the much reproduced photograph of "Negrito Archers" by Frances Benjamin Johnston, pictured here. (Johnston was another of the Fair's important female photographers.) Obviously posed, the menace in this shot, whether symbolically directed against animals or other humans, is also consciously deflected away from the camera and, hence, from the viewer looking through the eyes of the photographer. All of the subjects are looking down. So danger is symbolized, abstracted, and ritualized, preserving the sense of the primitive without threatening the viewer or the photographer. Any threat is contained, and the authenticity is obviously staged.[17]

Beyond the large number of photographs shot by professional photographers, there exist a number of private collections of photographs taken by amateurs and preserved at the Missouri Historical Society. Among these are collections that reveal the extraordinary range of interests of the photographer as well as the huge number of possible sites for taking pictures: from particular exhibits like the Simmons Hardware Company's Keen Kutter knife collection, to jewelry store kiosks, to Chinese porcelain, to Igorots playing with the photographer's camera. Almost any perspective lent itself to a photograph; consequently, the general collection reveals the extraordinary range of interests of photographers.[18]

FIGURE 3. "Negritos Shooting Bow and Arrows." It would seem that one of these fierce warriors, at the far right, is, in fact, a woman. Note that the arrows are pointed away from the camera and the viewer. (Frances Benjamin Johnson; Library of Congress Prints and Photographs.)

One common way to reproduce and share images of the Fair was through the picture postcard. Commonly seen at the Chicago World's Fair in 1893 and bolstered by the one-cent mailing price established by the Post Office after 1898, postcards could be purchased at the Fair or received as promotional items, or even made up from private photographs taken by individual visitors.[19] The format of these cards is usually very different from present-day commercial examples. For instance, many of the cards, even the commercial cards, had only limited, if any, space for a message; they consisted of the mailing portion on one side and a photograph that took up most of the other side. Subjects varied enormously in theme. There were many depicting the Philippine Villages. Others reproduced well-known photographers' work, such as Frances Johnston's "Negritos Shooting Bow and Arrows." On occasion, postcards showed the bare breasts of female denizens of the African villages. The Pike was also an immensely popular subject, as were staid and elegant pictures of the main buildings. There were even novelty postcards with transparent sections that would seem to light up if held against a window or a lamp, reproducing the remarkable nighttime illuminations. Almost immediately, these images became an object for collectors, and, in some cases, sets of postcards were sold exactly for this purpose. In fact, it is estimated that only half were actually sent through the mail, with many of the others placed immediately into scrapbooks.

Perhaps it is an exaggeration to say that some varieties of these early postcards represented a democratization of image making, but there certainly were possibilities for individuals to create a unique and inexpensive collection of photograph cards by transforming their own snapshots into a medium that could be shared widely with others. And what we see, in looking at the extant collections of some of these postcards, is, once again, an elaboration of the principal division between the ideology of the Fair reproduced from official photography in commercial cards and the individual input in amateur, or "real-photo," postcards.[20]

These real-photo postcards were produced in small commercial photographic establishments from snapshots and then cut to postcard size. While mechanically produced cards were sold to tourists, the real-photo postcards often depicted an acquaintance of the photographer or even the photographer herself, or an unposed situation.[21] Consequently, they were often without the usual narrative baggage of photographs that appeared in souvenir or guide books, although captions could be included by writing on the negative. Even if they were reproduced in a commercial photo shop, they represent an individual assertion about the importance of a particular display or aspect of the Fair. And, in some cases, such as the "Igorrote Dog Feast," we get a candid shot of the interaction between the watchers and the Filipino performers, revealing the quite remarkable close, even intrusive, proximity of the audience. Such shots reveal that patrons took the opportunity to wander around the native villages in the compound and talk to the inhabitants—something that photo books rarely suggest.

Although there were thousands of real-photo postcards, over a thousand different commercial postcards were issued and sold at the Fair. Many of them were touched up with color. The Samuel Cripples Envelope Co. of St. Louis held the official license to sell postcards, but there were other companies that also competed for the tourist dollar. Many of these postcards depicted official buildings and display palaces, but, as the historian of St. Louis Fair postcards has written, despite its lofty aspirations, "the exposition's stellar and most popular attraction was 'The Pike,' its fabulous mile long amusement area."[22]

Rarely noticed or consulted by historians, the stereograph was also an immensely popular form of photographic art at the Louisiana Purchase Exposition and probably more widely viewed than ordinary two-dimensional

FIGURE 4. "Triumphant Host of King Jesus: S. S. [Sunday School] Parade, St. Louis Day." Like many stereograph shots, this hugely busy perspective becomes clear and almost magical in the portrayal of three dimensions. (Library of Congress Prints and Photographs.)

productions. The era was noteworthy for the widespread production and purchase of stereographs. This parlor art remained a common fixture in a great many American homes into the 1920s. The stereograph constitutes a distinct form of photographic art, with its own aesthetic possibilities. Images appearing in this format from St. Louis give a very different impression of the Fair from the two-dimensional, flat, and elegant official views reproduced in books and promotional material. While two-dimensional photography lent itself to grand vistas, the motionless beauty of Beaux-Arts architecture and gardens, or the posed captivity of a framed scene of friends or curiosities, the implicit narrative of stereos often focused elsewhere. With their multiple planes and dimensions, they gave the impression of a layered, complex reality, in which the viewer could almost see around objects to something standing behind them. In fact, the stereo photographer often intentionally sought subjects that had an obvious foreground, middle ground, and background for just this reason. As a medium that celebrated complexity, stereographs were most appropriate for capturing the confusion of crowds or crowded spaces, rather than long-distance scenic views.

productions. The era was noteworthy for the widespread production and purchase of stereographs. This parlor art remained a common fixture in a great many American homes into the 1920s. The stereograph constitutes a distinct form of photographic art, with its own aesthetic possibilities. Images appearing in this format from St. Louis give a very different impression of the Fair from the two-dimensional, flat, and elegant official views reproduced in books and promotional material. While two-dimensional photography lent itself to grand vistas, the motionless beauty of Beaux-Arts architecture and gardens, or the posed captivity of a framed scene of friends or curiosities, the implicit narrative of stereos often focused elsewhere. With their multiple planes and dimensions, they gave the impression of a layered, complex reality, in which the viewer could almost see around objects to something standing behind them. In fact, the stereo photographer often intentionally sought subjects that had an obvious foreground, middle ground, and background for just this reason. As a medium that celebrated complexity, stereographs were most appropriate for capturing the confusion of crowds or crowded spaces, rather than long-distance scenic views.

For the purchaser, all that was necessary to use this medium was a viewer, a source of light, and stereographic cards—each consisting of two side-by-side shots of the same scene or image taken, as the eye would see them, through the slightly separated lenses of two different cameras. The result was a photographic art form in which every example strove for hyperrealism; each view emphasized the technology that produced three dimensions as much as it did the subject itself.

For its inventors, the stereograph was initially a laboratory instrument designed to illustrate and explain theories of vision and depth perception. But it quickly became a popular medium, especially with the invention of photography. During the late nineteenth century, the size of stereo cards became standardized to three and a half inches tall by seven inches wide, making it possible to manufacture universal stereographic viewers. One of the inventors of a classic handheld viewer was the American poet and physician Oliver Wendell Holmes. Calling the stereo a "mirror with a memory," he enthusiastically proposed standardization of the production of cards and viewers. Holmes ardently endorsed this new medium and rightly predicted its immense popular success. An art form that allowed the viewer to enter into the scene imaginatively, he noted, "To this charm of fidelity in the minutest details the stereoscope adds its astonishing illusion of solidity, and this completes the effect which so entrances the imagination." So true was the invention to revealing nature, that even a woman's age could not be hidden, he declared.[23]

The stereographic industry reached its height of popularity in the early years of the twentieth century. Consolidation in the industry meant that four or five companies shared most of the production and market. Boxed sets of stereo cards and viewers were frequently sold by itinerant salesmen. Underwood and Underwood, for example, employed as many as three thousand door-to-door agents over the summer, selling as many as twenty-five thousand cards a day in 1901. Sears and Montgomery Ward handled a large mail-order trade.[24] In many cases, companies included instructions, maps, and guides along with their boxed sets of scenes.

A contemporary instruction booklet published by Underwood and Underwood in 1904 suggests the way companies hoped the consumer would use the viewer and cards. Written by Rufus Rockwell Wilson, and accompanying "A Visit to Our National Capital," the brochure included instructions about how to enjoy the product as well as a short discussion of the psychology of looking through a stereoscope. The author advised first consulting the enclosed map of the city and locating the exact position of each photo about to be viewed. "Then," he instructed, "as you turn to the scene, think intently of your position in Washington, the direction in which you are looking and of your surroundings—the places of importance, not only in front of you, but to your right or left or behind you." Next comes his most remarkable and revealing suggestion: "It intensifies one's experience greatly to make hand-motions or to point, while keeping your head in the stereoscope, toward these places of interest."[25]

In his concluding remarks, Wilson turned to the mental effects of stereo viewing and its comparison with the experience of tourism, ostensibly quoting a psychologist who was an expert on such matters. "The essential thing for us is not that we have the actual physical place or object before us, as a tourist does," he suggested, "rather than a picture, but that we have some at least of the same facts of consciousness, ideas and emotions, in the presence of the picture that the tourist gains in the presence of the scene." This, he concluded, was "entirely possible in the stereoscope." The selling point, if not the actuality of the stereo, was its ability to transform the viewer into a virtual tourist, experiencing the joys, emotions, and edification of an actual trip or, perhaps in hindsight, by amending a real experience with new perceptions.[26]

World's fairs were particularly popular among consumers of stereographic sets (one manufacturer produced four thousand stereo views of the Philadelphia Centennial in 1876), in part because they could include several of the most important genres in the stereo repertoire. In addition, world's fairs offered the possibility of different shots and unusual perspectives. As one historian of stereography has noted, "stereo photography was marketed as a vicarious trip to points unfamiliar to most Americans." In this era of burgeoning tourism, images could stand in for a trip to the wonders of the American West, American cities, historic New England, and, of course, the grand tour to European cultural centers. In addition, stereo sets often depicted historic events, such as the building of the transcontinental railway, the Boer War, and the Johnstown Flood of 1889—just as the concessions on world's fair midways could reenact these events.[27] In its 1904, one-hundred-picture set of a trip around the nation (offered at $12.15), Underwood and Underwood suggested that a stereo might be even more valuable than tourism itself. "*Do not hurry*," it advised in its instruction booklet, "Tourists often lose half the meaning and half the pleasure of a journey because of their nervous way of scampering from one sight to another without stopping to think about what they see." A stereo allowed the individual to have a leisurely visit and not worry about train schedules, transportation, and other distractions.[28]

Underwood and Underwood also proclaimed that its products democratized tourism, bringing foreign places into the parlor of any family. This memorable experience was "possible for all sorts and conditions of men." The stereoscopes were truly, it concluded, "*windows through which you can see the real thing*" (their emphasis). If this inducement was insufficient, the brochure added that its tourist views promoted the excitement and awe of the experience. Thus, at Yosemite, you would "feel yourself actually there on that perilously narrow shelf where those surefooted horses are pausing; you almost hold your breath as you peer down into the bottom of the valley."[29]

An essential aspect of tourism was the encounter with different cultures and peoples. Thus, the subject of the stereo was often individuals or groups. These ranged from pornography or thinly disguised erotic views (an early favorite subject) to ethnic and racial types. There were frequent reproductions of Native Americans, as individuals or in villages or as displays at world's fairs. Inevitably in this time of explosive race relations, there were also frequent, often crude, depictions and stereotypes of black Americans.[30]

Given the versatility of the camera, despite some technological limitations at the time, the stereo acted as something of a combination newsreel, travel book, and nationalist cheerleader. As a record keeper of the Spanish-American War, the stereo enabled the viewer, in the words of cultural historian Edward Earle, "to participate in the visible effects of the age's political and economic conquests."[31] As a result of these depictions, the stereo became part of the cultural imagination of the nation. In the process, stereographs became stereotypes through their endless iteration and re-release. New publications often included older or previously used images. Like photographs and guidebooks, stereos contributed to the visual expectations of Americans as they encountered their nation and the rest of the world through travel or, more likely, through consumption of these visual materials.[32] As much as any popular contemporary medium, they generated the American landscape. As Samuel Batzli writes, the stereograph and the circulation of its repeated images trained Americans about what to expect when they finally saw a famous site and, indeed, instructed them what to look for. This, he argues, was particularly the case for national monuments and spectacular natural settings such as the Grand Canyon.[33]

Historians of the stereoscope have quite appropriately emphasized the setting of viewers—generally the parlor. This association with family and respectability through the frequent representation of female viewers in advertising copy was partly designed to erase whatever prurient reputation might have been associated with the pornographic possibilities of the medium. At the same time, because the parlor also functioned as the site of courtship, with the possibility of physical proximity, the stereo took on overtones of voyeurism and the erotic. In general, however, this medium played to the eagerness of Americans to comprehend the rapidly opening and expanding world around them. Thus, as Laura Schiavo writes, "The parlor stereoscope and its views enacted a confidence in vision and in the transparency between the object and its representation. They functioned as material evidence—consumable and repeatable signs—of abstract dreams of nineteenth-century imperial vision."[34]

The stereograph, in particular, participated in this imperial vision. Particularly after the Spanish-American War, companies that produced large numbers of photographs suggested in their written materials accompanying scenes of Asia, that it was the duty of the United States to "civilize" other peoples. Throughout the decades surrounding the Fair, stereotypes of Asia tended to repeat the same narrative clichés found elsewhere among American prejudices: beautiful, fascinating Japan and stagnant, dirty, and hostile China. In depicting racial typologies, stereos differed little from the displays of racial and cultural difference then in American ethnographic museums or exhibitions. It should be noted that extensive photographic records of the Philippines were already available before the Fair. Thus, many Americans might already have been familiar with the sorts of exhibits on display in St. Louis well before they arrived.[35]

If stereographs created expectations for future visitors to national monuments and to the world's fair in St. Louis by developing visual clichés about these spots as well as world travel, recent events, and the peoples of America and beyond, they also established a vision and memory of the Fair, even for those who did not make the expensive trip to hot and dusty St. Louis in the summer of 1904. Like all of the other souvenir productions of the Fair, including the two-dimensional photographs, guidebooks, and newspaper and journal articles, the stereo participated in this creation of meaning and memory of the Louisiana Purchase Exposition.

This meaning contained, at its core, racial and ethnic stereotypes that were already visually present in American culture, reaching even the respectable audience in parlors across the country. Thus, visitors to the Louisiana Purchase Exposition might bring with them considerable visual baggage derived from a variety of sources. Certainly the two-dimensional visual repertoire available in books, newspapers, and journal articles presented a Fair of amazing perspectives and possibilities. And the huge wave of prose advertising made it inevitable that many of the visitors already knew precisely what they wished to see. But the question remains, did the stereographs present a different picture of the Fair and the experience on the grounds? If we look closely at this visual record, will we see something different?

The most profound impression conveyed by the stereographic record from the Louisiana Purchase Exposition is one of crowds, a hurly-burly of exhibited items crammed cheek-by-jowl in display cases and kiosks, of parades and general movement. The stereograph, as an art form, works best in exploiting multiple planes of vision—generally about three—so each picture emphasizes not just what is depicted but the depicting itself—the medium. The best of these are advertisements for the visual miracle of three dimensions, showing not just reality but the idea of an enhanced reality. Looking at these shots, the viewer is first surprised at the ability of the camera to exaggerate the depths of vision. As millions of Americans did during and after the Fair in boxed sets or in recirculated images in other collections, they must have gained an impression of movement, people, confusion, and, perhaps most important, the amazing diversity of things to be seen. Thus, the focus, while directed at one object—a collection of oranges from California, for example—is always also on the possibility of seeing something else around the corner or behind it. Not unexpectedly, there are a great many shots of the Pike and hints of its possible pleasures. Indeed, if the two-dimensional photography published by the Fair administration sought to capture the still beauties of the Beaux-Arts center of the Fair with its gleaming display palaces and meandering visitors, the typical stereo focused on the Pike and its milling crowds. Appropriately, the Whiting View Co. entitled one of its views, "A Surging Sea of Humanity," and another, the "Swarming Pike." This was far from the intended message of unity and hierarchy articulated by the Fair planners.

An existing boxed set of Louisiana Purchase Exposition stereographs sold by Underwood and Underwood in 1904 contains about fifty or so stereo cards.

Almost all of them have elaborate and fanciful captions on the reverse of the card, with a shorter identification in six different languages. Although there may have been some initial prescribed order, there are no numbers on the cards, thus allowing the viewer to determine which he or she wishes to select. Furthermore, the whole set was placed into a double box designed to resemble two books, side by side, making it possible to mistake them for novels when stored on a bookshelf in the parlor.

Whether these views of the Fair were sold as boxed sets, individual cards, or traded or borrowed, the stereo had one important characteristic that published two-dimensional pictures generally lacked. Most of these latter were printed and bound in picture books, either as illustrations or in some cases as visual narratives of the Fair. As such they were immovable; they remained frozen in their published order. Stereos, on the other hand, despite the aggressive instructions generally accompanying them, could be viewed out of order, or selected by theme, or seen randomly, allowing the purchaser far more freedom to construct an imaginary storyline. If these were purchased at the Fair, they could also be used to recall and narrate experiences on the grounds—even creating the possibility of including elements that the visitor did not actually see in person.

Although there is certainly no way to count the number of stereos taken and sold by the ten companies that received official permission to photograph around the Fair, and no way to be sure that surviving examples are representative of the whole, the collections at the Library of Congress and the Missouri Historical Society in St. Louis give some distinct impressions about the favorite subject matter of photographers. They also seem to confirm the existence of a different aesthetic from the two-dimensional photo-illustrations contained in souvenir books.

A significant number of these stereos depicted parades through the fairgrounds. These were daily occurrences and a key element in the Fair experience. There are many reasons for this prominence. Various organizations, such as the suffragists or prohibitionists or Sunday school organizations, used the venue as a form of promotion. The Fair also advertised itself, mounting parades of representatives from concessions on the Pike, or its special ethnic displays of Native Americans, or representatives of the Philippine Villages. Given the relatively high price of visits to more than one concession, such frequent parades, particularly on the Pike, brought the exotic performers outdoors and into the view of those who did not or could not pay to enter.

Stereo shots of parades also confirm the large and impressive footprint of the military on the Fair, an impression supported by the daily program of

FIGURE 5. This bird's-eye view of a parade suggests both its length and extent around the loop and the large crowds that gathered to see it. (Library of Congress Prints and Photographs.)

events. From processions of veterans, to foreign nation marching bands like the French and English groups, to West Point cadets, one among many aspiring junior soldier squads who were encamped on the fairgrounds, to Filipino Scouts, and to representatives of local state militias, this military element was a constant visual presence. It served as a reminder both of recent military turmoil in Europe, the Boer War, and the Philippines Insurrection, as well as other past American wars, including the final and recent suppression of the western Native Americans. In total, there were around one hundred different military groups and over twenty thousand military men (not counting permanent guards) present over the course of the summer in St. Louis. The parade, with its emphasis upon orderly procession headed by brass marching bands, suggested that the multiplicity of peoples, projects, ideas, and displays could be contained within an idiom of a military performance.

Stereos of other parades show commemorations of official state days, visiting and local dignitaries, and, in particular, parades of ethnic peoples under such titles as "Children of All Nations Parade," a stereo produced both by Keystone View Company and the H. G. White Company. Parades of Native Americans were also widely photographed and available as stereos. Sometimes they were depicted not as conquered tribes but as mounted warriors in regal dress.

Taken as a whole, the lasting impression of these photographs is a vision of complexity and masses of people. By adding three-dimensional depth to the sights of the fairgrounds and its buildings, displays, and people, these stereo views confirm how visually crowded and confusing much of the area was, from huge crowds, to cramped display halls, to riotous decoration, to the sheer multiplicity of possibility. This visual profusion implies that the careful plans of the administration, from Munsterberg's hierarchy of knowledge to the tedious and elaborate notion of an encyclopedia of process and product, may have been invisible or at least compromised by overlying impressions of confusion. Stereos, it seems, suggest the inherent disorder in offering too many possibilities, by constructing a visual experience of the Fair on multiple planes. I will discuss in the next two chapters how visitors might have engaged that disorder.

CHAPTER FIVE

MRS. WILKINS DANCES

The comparison I have drawn between two-dimensional photography and stereography might seem to imply that ordinary photographs lack depth of meaning or gain usefulness only as illustrations rather than for their intrinsic qualities. Nothing could be further from the truth. If the foregoing suggested that, as genres, these two varieties of photography opened up the subjects they depicted to different sorts of readings, and were generally employed to create distinct impressions of the Fair, there are other ways to understand two-dimensional photography, to inquire more deeply into its significance and the meaning it offers. As Siegfried Kracauer describes it, the photographer "resembles perhaps most of all the imaginative reader intent on studying and deciphering an elusive text." So too must the imaginative historian be a reader who recognizes that the frame of the photo is only provisional and that meaning depends upon the context in which it is read. The photo's meaning, he continues, always "refers to other contents outside that frame."[1] Concentrating attention on the interior of the photograph as well as retrieving its contexts can greatly enhance our understanding of the cultural issues that were at stake on the fairgrounds of the Louisiana Purchase Exposition of 1904. This can open a window onto the preconceptions that visitors might have brought to the Fair as well as suggest how they experienced its wonders. If circumstance and background constitute a significant portion of meaning, then the task of the historian is to conduct an open-ended exploration of such possibilities, to view photography as documentation that both reflects and reflects upon its cultural contexts.[2]

Roland Barthes has written that a photograph may contain multiple signs, which can be read according to different systems of meaning. As he puts it, the reading of a photograph is "always historical; it depends on the reader's 'knowledge' just as though it were a matter of a real language."[3] In this chapter I will focus on one particular photograph that suggests multiple readings, or "contexts" as I call them. My intention is to explore the way that one particular and suggestive image can serve as a summary of symbolic meanings, of possible cultural significances that impinge upon this single shot but, at the same time, reveal the photograph to be a significant comment on the Fair and on the times.

Certainly one of the most unusual and intriguing photographs from the Louisiana Purchase Exposition is "Mrs. Wilkins, Teaching an Igorrote-Boy the Cake Walk," taken by the legendary Jessie Tarbox Beals. Aside from its unusual content, it is one of several images given a playful caption by the photographer herself. In some shots, this caption was written directly onto the negative by the photographer; in this case, it appears to have been added later.[4] Fortunately, there are companion shots as well as a larger context of other photographs that indicate its place in Beals's work as well as the photo record of the Exposition. Beals, the Gerhardt sisters, and many of the other prominent photographers of the Fair often concentrated on capturing "portrait-types," the generic term for individual or group images that presume to stand as the symbolic representation of a whole people and that are discussed in the last chapter. Thus, a picture of a Native American is both individual, unique, and intended to represent a type, as is the single or group picture of Filipinos, for example. We need to recall, here, that this format is one of three types picturing peoples of the Fair. Much of Beals's most interesting work belongs in this category.

Portrait-types are generally a studio portrait of a single individual, usually sporting elaborate native dress or a cultural implement of one sort or another. Interaction between audience and native peoples was another frequent subject. These shots depict audiences gazing at natives in various circumstances: at work, eating, drinking, dancing, or simply standing as an object of observation. This photo type suggests the intrusiveness of the visitor, recording the close quarters and visual intimacy between subject and object. Often in these pictures, there is a vivid contrast of dress between the near nakedness of the native in his or her "natural" surroundings and the highly reserved, almost formal, attire of the visitor, who is easily identified as an outsider. The third exemplar captures some interaction between audience and the subject

FIGURE 6. Type 1: In full regalia, this Native American stares directly into the camera and at the viewer. (Gerhard sisters; Missouri Historical Society, Photographs and Prints Collections.)

of interest; the viewer and the observed may be revealed in a conversation or some other form of mutual intercourse that seems to break through cultural differences or at least capture an interaction.

Even among interactive photographs in this last category, "Mrs. Wilkins, Teaching an Igorrote-Boy the Cake Walk" was a remarkable image for 1904 and an anomaly in the existing visual source material of the Fair, partly because of its intriguing subject matter but especially for its suggestive caption. Contrast this portrayal of the interaction of Filipinos and St. Louis audiences with the visual material provided in 1973 by Margaret Johnson Witherspoon

FIGURE 7. Type 2: These Igorot men and boys huddled together while fairgoers passed through their village in the Philippine reservation. (Jessie Tarbox Beals; Missouri Historical Society, Photographs and Prints Collections.)

in her short memoir of the Fair. This modern author included two pages of pen and ink caricatures of "primitives" under the title "Stylish Ladies on Way to View Naked Savages." If this contrast of ladies and savages, fashion and nudity captures the pervasive atmosphere of erotic interest in the exotic cultures brought to St. Louis in 1904, and seen as a titillating episode even seventy-five years after the Fair, then Jessie Tarbox Beals's photograph of Mrs. Wilkins is that much more revealing for reversing this prurient interest and its stereotypes about race and civilization.[5]

Much can be learned from a preliminary deconstruction of the elements of the Wilkins photograph. In the first place, the caption is unusually informative, indicating the sort of dance being performed and the identities of the performers. As Roland Barthes suggests, the linguistic elements surrounding an image are attempts to anchor or fix meaning by emphasizing certain signs within the image.[6] Thus, the caption directs the attention of the viewer to Mrs. Wilkins. She is the teacher, the initiator of the action. We are asked to view the dance through her eyes and intentions.

This linguistic focus represses other possible interpretations of the pho-

FIGURE 8. Type 3: "Igorrote Singing Lessons." This photograph shows the interaction between observer and observed. Perhaps the teacher is instructing these Filipino boys in American tunes. Patrons elsewhere on the fairgrounds were delighted when Philippine natives played patriotic songs. Part of the *mission civilatrice* was teaching Western culture. (Jessie Tarbox Beals; Missouri Historical Society, Photographs and Prints Collections.)

tograph. For example, it is apparent that the Igorot boy is taking the lead, as the male partner would do in such a dance. It is also true that he has been dressed up for the part, wearing a top hat of sorts (characteristic of the overdressed finery worn by professional and amateur dancers of the cakewalk). But this addition of top hat and cane only emphasizes the boy's nudity. When Beals and Mrs. Wilkins posed this shot, they clearly had no interest in mask-

FIGURE 9. "Mrs. Wilkins, Teaching an Igorrote-Boy the Cake Walk." (Jessie Tarbox Beals; sold to the Louisiana Purchase Exposition Company; Missouri Historical Society, Photographs and Prints Collections.)

ing the Igorot's body. They only provided minimal apparel to symbolize the dance: a hat and cane. It would be entirely legitimate and fascinating to pursue the boy's identity. He is perhaps one of the figures in the next photograph ("Mrs. Wilkins Sings"). We might also speculate about the meaning of his obvious pleasure. Was it with the dance itself or the symbolism of the racial separations that he was bridging? Did he enjoy the adult male (lead) role that the dance bestowed upon him?

It is my purpose in this chapter to focus, as the caption instructs us to do, on Mrs. Wilkins and her lesson. Above all, the photograph records an action, and, much as we are drawn to it as a text filled with symbolism, it is primarily important as a document recording an action. Although we cannot

FIGURE 10. Mrs. Wilkins sings. The boy with the short, striped loincloth may be her dancing partner. (Jessie Tarbox Beals Collection, Schlesinger Library, Radcliffe Institute, Harvard University.)

be absolutely certain that Beals herself etched the caption onto the negative, it is likely that she did so as a way of explaining what might otherwise be unintelligible. The photograph itself was part of a series sold to the Louisiana Purchase Exposition Company. But it was (apparently) not published until Eric Breitbart's 1997 photographic collection of Fair photographs. Beals did not publish her collection of photographs from the Fair independently.[7]

The photograph is set within the Igorot Village, in the midst of the large Philippine Reservation, with native structures visible in the background; the caption identifies Mrs. Wilkins as an individual. Her partner, however, is accorded only a generic name and gender.[8] This implication of social distance and superiority and the subject-object relationship was common between spectator and spectacle throughout the Fair. But contrasted with other "teaching" photographs taken at the reservation, showing American women instructing natives, this has remarkable and unexpected elements. Mrs. Wilkins is engaging in the much-praised mission of American civilizers: she is giving a lesson. In a companion shot, she is pictured instructing Igorot children in singing, although we do not know what songs. And the dance, particularly the popular cakewalk, was probably not what was in the minds of most missionaries of civilization in 1904.

In the dance photograph, the lesson is plainly subversive to the project of Americanization and uplift and, perhaps, even scandalous, even if her partner is identified as a "boy" and not a man. Instead of instructing the Filipino in his new civic duties as a colonial subject, such as how to salute the American flag properly or how to read and write English, as the colonial schools taught, she gives a lesson in a popular dance step whose origins clearly define it as a derivative of African American culture.

This chapter will explore Beals's intriguing photograph of Mrs. Wilkins in the contexts within which her action might have been seen at the time as well as how historians might now interpret it. Looking across these two different time periods and two modes of interpretation, this technique of contextualization explores ideas that viewers might have brought with them to the Philippine Reservation as well as what the modern observer, looking back on this period, might conclude, especially about the operation of notions of race relations at play in 1904. On occasion, this dual perspective will require us to return to interpret certain elements of the photograph itself. It will also consider intellectual and cultural sources that were part of the thick context of ideas about race in 1904 that hung over the Fair.

Laura Wexler has written of Beals that her "pictures extended sympathy, but they also fully eradicated any traces of the military surveillance that had accompanied the Filipinos to St. Louis." (Here the author is referring to the just-concluded brutal suppression of the Philippine independence movement.) Wexler places the photographer within a larger cohort of female photographers who, she argues, "sentimentalized the view of imperialism." Their "innocent eye" denied the violence, warfare, and force that accompanied the imposition of imperialism on the archipelago after the Spanish-American War, constructing, instead, a "domestic vision."[9]

Beals's other photographs of Filipinos ranged from stereotypic portrayals to depictions of humorous reversals. For example, she shot nude Igorots as well as natives in western dress. She captured Igorots butchering a dog for dinner as well as killing a chicken for a wedding feast. One of the most interesting shots depicts a much-photographed Igorot boy (Singwa) standing behind a camera, evidently taking a picture of the photographer herself.

In the broader context of colonial photography popular at the time, St. Louis presented an unusual photographic opportunity, because the colonial peoples were brought to the United States and placed in a zoolike setting. Photographs taken there joined many others taken on location in Africa and Asia in circulation at the time. Among the most popular of these were stereograph sets and postcards featuring photographs of aboriginal peoples or scenes from the Boer War.[10] So too were portraits of Native Americans, pictured, as the cliché had it, as a proud, once fierce, and now dying race. As Anne Maxwell notes, this vision was part of a larger culture of depicting captive societies, common in Europe as well as the United States in this era and of special importance at world's fairs. Indeed, one of the first large living tableaux colonial displays of subjugated peoples appeared in Paris in the 1880s. "Authentic" models of villages reappeared at the Exposition Universelle, also in Paris in 1889, and thereafter at Chicago and St. Louis. These were built to satisfy observers as much as they were to reproduce actual dwellings and communities. Even so, Maxwell noted, there was a difference in the reception of these displays according to the observer's intent. Scientists, she suggests, were more interested in discovering an objective vision of culture and so emphasized race and civilization, while tourists were more likely attracted because of sensational advertising. On occasion, she notes, such photographers as the Gerhard sisters reversed this colonial gaze with their sympathetic, unsentimental photographs of famous figures like Geronimo at St. Louis. And if colonial peoples seemed to prefer studio photographs like those of the Gerhards, for which they could exercise some control of their image, making eye contact with the lens and the viewer behind the camera, tourists and anthropologists apparently preferred seeing them in their "native" contexts and unengaged with the audience.[11]

If native dress in a full-length shot or a close-up facial study characterized portrait-type photography, then introducing an audience into the background profoundly shifted the meaning of this genre. But there is something peculiar about the constitution of this audience in St. Louis. Where they might be expected, one notable group is rarely if ever in attendance in any of these visual records. African Americans almost never appear as part of audiences, anywhere in the photographic record of the Fair. One historian estimates that about 1 percent of the visitors were African Americans.[12]

Initially, there had been considerable local enthusiasm among the black elite about the economic and social possibilities of the Fair. The largest black newspaper in the city, the *Palladium*, predicted a singular occasion to display the progress and status of the community as well as provide for new business opportunities. Some jobs associated with construction and maintenance certainly did materialize, and the Fair attracted a number of prominent ragtime pianists to the city. The paper faithfully reported the names of African American visitors from out of town. But eventually, this enthusiasm diminished and most of the activity that mattered to the community occurred in businesses outside the fairgrounds.[13]

Black Americans generally and correctly decided they were unwelcome. Not only did they feel unwanted because of segregated restaurants on the Pike. There were other affronts such as the planned encampment on the grounds of African American soldiers (the Eighth Illinois Regiment), which was rejected, while scores of white units were encouraged to visit. A projected Negro Day at the World's Fair, either in August or September, which Booker T. Washington was to address, was cancelled by the participants themselves. Nor would the administration accept a "Bureau for Colored People" detailing the accomplishments of black Americans, although at one time the Louisiana Purchase Exposition Company considered building a "Colored People's" center, even going so far as to approve a design for the project. Although David Francis assured African American patrons that they were welcome on the grounds, there was significant negative publicity in the St. Louis newspapers that discouraged patrons of color. And with good reason: in a letter to Concessions Head Norris Gregg at the end of July, a restaurant owner explained his policy. He would serve all comers regardless of race or color, "only we serve Negroes on another part of the premises."[14]

There were consequently no displays of progress since the Civil War, no portraits of significant black Americans. African Americans did not make up the audience in significant numbers. Nor, of course, did they appear in the anthropological hierarchies constructed by W. C. McGee except by implication. Generally, dark-skinned peoples—such as Pygmies or the "dying" Negritos from the Philippines—were recruited by the Fair management as oddities. Neither wards of the state like Native Americans nor conquered peoples hurried on their way to civilization and progress like the Filipinos, black Americans had no place in the audience or ideology of the Fair. Their absence, indeed, was glaringly conspicuous. In part, this nonappearance may be explained by the small local St. Louis black population and the corresponding slight population throughout the state. But there is no doubt that formal and informal exclusion whitened the face of the audience. And this absence, in turn, may have affected the racial and ethnic interactions that frequently occurred on the fairgrounds.[15]

This had not always been true at other fairs. For example, at the Paris Exposition of 1900, W. E. B. Du Bois and Thomas J. Calloway collected and mounted a photographic display of "Types of American Negroes" and "Negro Life in Georgia." As Du Bois explained, these were "several volumes of photographs of typical Negro faces, which hardly square with conventional

FIGURE 11. "The Extremes Meet—Civilized and Savage Watching Life Savers' Exhibition, Igorote Family at World's Fair." This rare and unusual shot shows Filipinos in the parade audience as well as what looks like one black man. (Library of Congress Prints and Photographs.)

American ideas."[16] In fact, what Du Bois accomplished in this collection of intellectuals, doctors, middle-class club women and businessmen—a thoroughly middle-class and perfectly respectable collection—was to reverse and subvert the notion of the native portrait-type, transforming it into its reverse, the photograph of a dignified individual usually depicted in a parlor or office setting. This distinguished group contrasted sharply with contemporary visual caricatures of blacks particularly rampant in the United States and Britain, where the term *Zulu* referred to both natives of Africa and black or black-faced entertainers and was also a generic term for exotic peoples displayed in dime museums and other popular culture entertainment venues in the late nineteenth century. To counter such spurious scientific displays and racial portrayals and impersonations in these cheap spectacles for exhibiting the exotic was undoubtedly something Du Bois had in mind for the Paris World's Fair.[17]

In a few other instances, African Americans had found their own voices at fairs. Black representatives at the Charleston Exposition in 1901–1902 had succeeded in establishing their presence and even sponsored a Negro Building, "the pride of the Negro men and women of this city." The structure contained displays from the Hampton Institute and Tuskegee Institute such as samples of fine needlework, elegant furniture, and manufactured items. As one enthusiastic observer put it, vainly, it turns out, "It is certain that future exhibitions of the first class in America will never, in the future, be without its Negro Department. The mistake of Chicago [World's Columbian Exposition, 1893] will not be repeated in St. Louis."[18] But it was. Aside from the assurance that black Americans would not disrupt or challenge the racial charts and graphs of anthropologists, the effect of this exclusion was to create a largely white audience. Perhaps that made it easier for Mrs. Wilkins to perform her dance, for she could do so without overtly transgressing American racial etiquette before a mixed audience where everyone would have been reminded of the trespass.

Her action was even more controversial because of the reputation of the Igorot peoples. Visitors to the Philippine Reservation knew what to expect well in advance of a trip to the Fair. In fact, this was the third exposition at which Filipinos had been exhibited (Spain in 1887 and the United States in 1901), although nowhere yet on such a grand scale as St. Louis.[19] Guidebooks and picture books generally marked this ethnic group as special. But it was the press that printed the lurid stories of Igorot feasts and male nudity that drew huge crowds to the site. Hundreds of stories appeared in American newspapers during the summer of 1904 that focused on the Philippines and Filipinos. While there is no possibility or utility in counting all of these, it is still evident that a great many, if not most, highlighted the strangest elements of the exhibit. Diet was probably the most repeated news subject, but there were many stories about skimpy clothing and nudity, descriptions emphasizing physical peculiarities, discussions of unusual customs, reports about ethnic diversity, even suggestions of cannibalism and human sacrifice. There were also frequent stories about possibilities for Filipinos to advance culturally. (That was Mrs. Wilkins's ostensible purpose in teaching.) Some stories related the exhibit to the ongoing political debate over the fate of the colony. But McGee's theories about hierarchies of race and civilization were almost never mentioned or examined. Instead, news became news because it was sensational and unusual. And it was almost always written in the prevailing American vernacular of racist attitudes and common biases against darker peoples.[20]

Clearly not the intent of McGee or any of the Fair administration, this attraction to the sensational and lurid disrupted the message intended by the larger exhibit with its justification of imperialism and its benefits for colonial peoples. This was apparent in an interview with Mayor Carter Harrison of Chicago. The *Chicago Daily* quoted him as saying, "They showed me the Igorrote. They are good looking men physically, but deficient mentally. Their chief achievement is to cut off the heads of their enemies. Their chief culinary delicacy is stewed dog." Then, wading abruptly into an argument about whether Filipinos could ever become Americans, he concluded, "No one would want them as American citizens."[21] Another example was an extensive article for the *Atlanta Constitution* about midway through the summer of 1904. Henry Grady, editor of the newspaper, explored the "startling Exhibit" of Filipinos at the Fair. The real purpose of the exhibit, he noted, was to justify the expenditure of $20,000,000 for the archipelago and the sacrifice of several thousand American soldiers in suppressing the Philippine independence movement. But, he noted, the audience at the Fair seemed far more interested in the "strange savages" than it did in the agriculture and educational exhibits from the Philippines. As he put it, the Filipino natives were the center of public interest, and there was a "never-ending throng of curious visitors watching every move of the little brown people, from the dress parade of the scouts down to the horrible dog feast of the savage Igorrottes."[22]

Two controversies concerning diet and dress only increased the scandalous reputation of the Igorots. The first developed when St. Louis newspapers published lurid stories about Filipinos sneaking off the fairgrounds to round up stray pets for their suppers. Their scant clothing also became an issue, particularly to the Board of Lady Managers, and reached the point where President Theodore Roosevelt himself had to intervene to prevent the denizens of the compound from being dressed up in more modest, American-style long pants. This incident proved to be a popular story and continuing source of humor for journalists who, for example, with tongue in cheek, carefully inquired whether a troop of representatives from the exhibit visiting the White House would wear loincloths or American-style dress.[23]

If the expectations of visitors to the Philippine Reservation were heightened by newspaper tales of the lurid, the bizarre, and shocking bodily display, that was obviously not what McGee had in mind when he laid out his elaborate justifications for racial and cultural hierarchies. In fact, it was to correct such popular false impressions, to impose a "scientific" anthropological context on this display of conquered peoples, to provide a reasonable argument about the universal benefits of imperialism that guided every step of his planning and presentation. Yet even the anthropologist himself was not immune to showmanship that gave anthropology the appearance of spectacle, for he delighted in engineering visual contrasts and juxtapositions. Thus, he arranged to place "representatives of all the world's races, ranging from smallest pygmies to the most gigantic peoples, from the darkest blacks to the dominant whites, and from the lowest known culture (the dawn of the Stone Age) to its highest culmination" side by side in the exhibit area.[24]

Writing shortly after the Fair closed, McGee rhapsodized over the remarkable accomplishments of the anthropological section. He insisted upon interpreting the exhibits as an important educational event to which visitors were guided uniquely by scientific curiosity. Indeed, he implied that visitors from far and wide, almost every state and several foreign countries, came to the Fair just "for the special purpose of seeing the African pygmies, the Ainu [a 'white' Japanese group], the Filipinos, the Patagonians, or the assemblage of North American tribes." The purpose was not to gratify curiosity or stoke wonder, he argued, but to satisfy the "intelligent observer that there is a course of progress running from lower to higher humanity." So impressed was he, that he boasted that "if another million dollars had been added to the Department of Anthropology there would have been three million additional admissions.[25]

McGee's other purpose was to employ university-based anthropology to correct and amend the long and well-developed commercial tradition of displaying human beings as curiosities and specimens in nonscientific circumstances. Anthropology, through its scientific measures, could do what the eye could not: precisely define and categorize people. Popular contemporary shows had also invoked science and sometimes made a half-hearted effort to place the native in his "natural" environment—generally in order to stake a claim for authenticity and underscore visual differences. Even in St. Louis, the display of humans was a crucial part of the broader effort to attract an audience—something McGee himself admitted. Once they were present, however, he hoped to instill two fundamental ideas into the audience. The first stressed the practicality and significance of the new science of anthropology itself. This promotion of a new field that had yet to achieve significant prestige in the academy proclaimed its immense practical value to a world increasingly unified by commerce and empire. So when McGee ambitiously claimed, "Anthropology is the science of man," he proposed it as the measure of humanity in its organic, linguistic, social, and productive dimensions. Even more grandly, he summarized his views as a compound of two distinct anthropological theories that reinforced each other's wisdom: the first stressed a color-coded racial hierarchy of five shades, and the second described ascending levels and degrees of civilization and culture. As the world's population was "classed by race and culture jointly, it was soon seen that the types of culture really represent grades or stages in development." Thus, physical evolution was extended by culturally induced evolution, leading to the great disparities that could be noted in contemporary history.[26] Race and cultural achievement (or lack of it) walked hand in hand in this scheme. As he enunciated this idea, "The whole Fair in its largest aspect is an exploitation of the Caucasian." Using the word *exploit* in its nineteenth-century sense, he meant that Caucasians were responsible for creating the exhibits in all the grand display palaces, in agriculture, machinery, liberal arts, and transportation, while other races were confined to specialized enclaves on the grounds.[27]

The assertion of extreme contrasts between subjecting Americans and subjected peoples in the suppression of Filipino independence has convinced one historian to argue that the Philippine War was instrumental in what he called "the manufacture of Caucasians." Thus, he notes, when the national press stressed the yawning differences in culture between dark-skinned "savage" Filipinos and civilized Americans, it simultaneously encouraged immigrants coming from Eastern and Southern European nations, who were designated as foreign to "Anglo-Saxon" stock, to think of themselves now as "white."[28]

McGee's final argument highlighted the inevitable and rapid acculturation and progress unleashed by the conquest. Placing Native Americans under benevolent American rule after the Louisiana Purchase from Napoleon in 1804 had allowed the "purposeful means of promoting the common weal," he said, as "aboriginal landholders" made their way toward U.S. citizenship. This process was vividly illustrated in the "typical Indian school forming the most conspicuous feature of the department." The same techniques of a largely industrial education were demonstrated in the Philippine Reservation. Native villages were arranged in ascending order of race and cultural progress, capped with a demonstration of American efforts at inducing general progress through a model school. As McGee put it in the journal *The World's Work*, "It presents the race narrative of odd peoples who mark time while the world advances, and of savages made, by American methods, into civilized workers."[29]

McGee, together with the chief of the Department of Physical Culture, made one other final gesture to illustrate the validity and usefulness of physical anthropology for measuring racial hierarchy and progress in civilization. This was the two-day event called Anthropology Days. It was held on August 12 and 13 as a sideshow to the International Olympics that took place on the fairgrounds and at nearby Washington University for five days later that month. (The Exposition Universelle of Paris, 1900, hosted the Second Olympic Competitions.) As the report from the Department of Anthropology noted, the two men frequently discussed the purported athletic abilities of "several savage tribes." Because of the "startling rumors and statements that were made in relation to the speed, stamina and strength of each," it was decided to hold the two-day event to test this supposed native athletic prowess. The result was an "utter lack of athletic ability on the part of the savages," although McGee believed that with training they might do better. His serious point, however, was to demonstrate that non-Caucasian peoples were wanting in every respect, not just culture and civilization but even physical skill and strength.[30]

Contrasting with the displays outdoors in the villages and other exhibits of peoples "untouched by the march of progress" were indoor representations of the way ancient civilizations had developed from the first use of tools to the discovery of the wheel and of iron. There were artifacts from Egypt and from Central American civilizations. These museum exhibits of historical anthropology were intended to confirm the same story of progress and evolution demonstrated by the outdoor villages in which contemporary peoples were spread out according to different degrees of progress.[31]

American scientific publications, such as *Science* and *Scientific American*, ran several articles in 1904 discussing anthropology at the Fair, and they largely repeated this message. Indeed, *Scientific American* explicitly linked the living tableaux to the fundamental message of the Philippine exhibit. As one writer put it, "The last doubt was removed from his mind that, in this matter of colonization, the latest and most difficult national enterprise upon which the nation has embarked, the government would achieve one of the most successful and beneficent works in the history of the United States."[32] Imperialism appeared convincingly just and benevolent.

Historians have rightly focused on this massive and complex exhibit as a centerpiece of the Fair; indeed, for McGee it was the beginning point and the culmination of everything else on the grounds. The entire exhibition could be interpreted to underscore the same points about race and progress. How much of this Mrs. Wilkins might have seen, understood, or remembered is not something we can know even though these ideas were ubiquitous in publicity about the Fair. Was her dance instruction her version of the American *mission civilatrice*? In fact, we do not know exactly why she was attracted to the Igorot Village, although there is some evidence that she was a school teacher. But perhaps by extrapolating from the general behavior of the audience, we can imagine something about her motivation. Admission receipts for the villages inside the Philippine Reservation reveal some interesting possibilities. We can determine, for example, what were the most popular of the villages and also whether or not most visitors actually entered the exhibits in their prescribed order, moving from savagery to civilization to grasp the message that McGee and others were so anxious to impart.

Given the entrance fee of $.25 for adults, it is possible to calculate that there were slightly over one million visits to at least one village in the reservation. This would mean that somewhat less than one-quarter of all visitors to the Fair (using my smaller estimate of general attendance) entered at least one village. However, if these visitors went to two or three, the percent of individuals who saw any one of the exhibits declines to one-eighth or even less. So we cannot be sure that most fairgoers had any direct experience with these villages. There is more to be learned from attendance records. Looking at total admissions from the five enclaves, the Igorot, Visayan, Moro, Negrito, and Bagobo villages, there is no question which was the favorite. By a wide margin, the Igorot Village had revenues that exceeded the combined total of all the other villages.[33] And there is also no doubt about what attracted audiences to this site: the diet (which was the subject of so much ribaldry and feigned horror in the American press) and the nudity of the men (*National Geographic* featured bare-breasted Filipino women in 1903). This strongly suggests that those who attended the Fair did not perceive, or care much about learning, McGee's complex racial geography, because they had other interests and expectations.[34] Of course, it was entirely possible for visitors to see other, free exhibits and museum displays in the Philippine Reservation that made the essential points about evolution and race, without trespassing the confines of the villages or experiencing their exotic cultural performances.[35]

All of these impressions, official and otherwise, about the Philippine Reservation constituted the immediate context for Mrs. Wilkins's adventure when she decided to dance with an Igorot boy and teach him the cakewalk. But there is another setting in which we must view this photograph. Because of the extremely potent racial discourse explicit in the Fair and its self-advertisements and because of the Igorots' dark skin, issues of racial attitudes and etiquette toward African Americans also formed an implicit milieu for her willingness to reach (literally) across racial boundaries. Although denied either a historical or anthropological presence on the fairgrounds, black Americans were omnipresent in another respect, for prevailing American racial attitudes challenged the complicated anthropological order preached by McGee. The audience certainly brought these attitudes with them to the Fair and may well have understood the different looks and culture of Filipinos in these terms.

The easy slippage between the darkness of Igorots and of African Americans in prevailing American racial attitudes figures clearly in the newspaper squib "How a Monkey Got In," published (and often republished) during the summer of 1904. The story recounts the failure of Dr. T. K. Hunt, the American governor of a province of the Philippines, and fellow inspectors to detect a pet monkey brought to the Fair by a member of the Visayan group.

FIGURE 12. Dr. Hunt of the St. Louis Pound serves stewed dog to David Francis while Igorots, portrayed as black Africans, look on. Bert Cobb, the artist, was a well-known painter of dogs. (Missouri Historical Society, Photographs and Prints Collections.)

A woman carried the monkey with her at her breast, "and the inspectors thought it was a Filipino baby."[36] The humor in this story lies in the assumption that the audience will get the joke and understand the plausible confusion of American officials who cannot tell a monkey from a native.

This same presumed similarity between simian and African led to the temporary display of Ota Benga in the Bronx Zoo. In other words, the culture of racism affirmed a visual connection between humans and animals that could be exploited as a joke. This comparison was the basis on which the Reverend Dr. S. P. Verner, who brought the Pygmy group to St. Louis for display, planned a lecture tour to England after the Fair. He hoped to bring along apes, a crocodile, and other specimens for viewing, but, as he put it, "A pygmy side by side with a chimpanzee would not be a poor [drawing] card, I am sanguine to think." In an earlier letter to McGee, Verner had speculated about the origins of one of the groups he would be visiting, speculating that they might be the missing link between apes or "a new species of anthropos or ape, or a combination, or at least the very lowest form of man on the globe."[37]

The 1890s and the first decade of the twentieth century were particularly trying and dangerous times for black Americans. Race consciousness and segregation, whose presence the Fair administration denied, was in the real world remarkably strong. There were two issues that most preoccupied black Americans and black organizations. Had their representatives met officially on the Fairgrounds there is no doubt they would have protested against the expansion of formal segregation and rampant lynching throughout the South and in border states like Missouri. Nationally, the rate of lynching peaked in the 1890s, with close to 140 (mostly black) in the worst year. Although the absolute number fell back in the early twentieth century, this vicious execution vigilantism continued at a high rate. Although sexual, or imagined sexual, transgressions were a frequent cause, most lynching was apparently as much about enforcing status, or about assault, murder, theft, economic crimes, or disrespectful language. In other words, lynching was a brutal and unofficial enforcement of racial separation and superiority for which segregation—with its legislated separate schools, stores, theaters, and transportation—was the formal version. In tandem, lynching and segregation enforced the rigid codes of racial separation that governed the South and most border-states.[38] So important and public were these inflictions of ultimate justice that, where they were frequent, a culture of spectacle grew up around them. This even included popular real-photo postcards shot by individuals and processed for collecting or mailing.

During this period, lynching had also occurred in Missouri, although on a more limited scale, reaching a height in 1898 (five), with two in 1903, none in 1904, and one in 1905. There were seven public murders of white men for the whole period 1894 to 1910, but the large majority during the time were black. In fact, however, unlike the South, the black population of Missouri was quite small, only about 5 percent of the state population in 1910 and a little more than 6 percent of the city of St. Louis at that point.[39]

On his famous trip through the South, published as *Following the Color Line*, Progressive reformer Ray Stannard Baker noted a peculiar ambiguity about prevailing segregation which suggests the existence of a third realm of more informal prohibitions. "The color line is drawn," he wrote, "but neither race knows just where it is."[40] Ever changing, shifting over time, and varying by place and custom, segregation achieved legal status through the passage of Jim Crow and miscegenation laws and, most important, by the Supreme Court case *Plessy v Ferguson* (1896). But beyond this tapestry of legal separations, race relations in the United States were also guided by a racial etiquette that informally determined the social and cultural interactions of black and white. In this realm we may discover the most radical implications of Mrs. Wilkins's dance.

The etiquette of racial relations, sometimes unspoken but widely understood, had as its purpose the assertion of white superiority, including both social and economic control, but with a minimum of overt force. As one historian has written, the rules of racial etiquette were designed to maintain separation: "The rule is that shaking hands, walking together, or otherwise associating in public, except on terms of superior and inferior, are not done." In the diligent search for distinctions, this extended in some cases to courts maintaining two Bibles for swearing-in witnesses: one for blacks and the other for whites.[41]

Perhaps the most poignant story ever written about American racial etiquette is W. E. B. Du Bois's "Of the Coming of John," published in 1903 as part of *The Souls of Black Folk*. This narrative, about an educated young black man's dangerous violation of etiquette, ends with his brutal lynching. The most remarkable insight of the story is Du Bois's recognition of the subtle linkage between the informal manners of separation and the violence that lay coiled beneath them, ready to spring into any serious breach.[42]

Mrs. Wilkins's dance takes on extraordinary significance precisely because of the number of these etiquette rules that it breached. This is all the more important because, as McGee so aptly characterized the Fair, it was a Caucasian event in which the audience, assumed to be white, could learn about the cultures and limitations of the peoples of other races. , McGee hoped that gazing at colonial villages, Native Americans, exotic others from the Orient, and the extensive displays at the Philippine Reservation would transform the polyglot ethnic peoples of America, not just Anglo-Saxons, but including Irish, German, and other immigrant stocks, into a Caucasian "us" looking at a primitive "them."[43] If we understand the gradual tendency in the United States to erase the conflation of ethnicity with race, the end of disparaging characterizations of Irish, Italian, and Jewish immigrants that identified them as separate and inferior races, then defining broad racial divisions between Caucasian and other races at the St. Louis World's Fair based upon levels of civilization was an important step along this direction of a more capacious definition of whiteness.[44]

This broad racial division echoed a more abstract intellectual distinction that pervaded the catalog of exhibits and the placement of displays. The Exposition had two unequal parts, defined by two grand categories of knowledge and perspective. The Caucasian or civilized world on display in the grand palaces was defined by historical knowledge that recorded progress as well as important past events. Indeed, the Fair itself was a celebration of American history. As Albert Jenks put it in another context, "the historic world looked upon" the savage world in the Philippines. This "primitive" world could best be understood through the human science of anthropology, which studied unchanging cultures that had neither past nor any future unless they could be somehow regenerated by European and American examples.[45] European nations, the United States, and Japan were all driven by development and change, and their exhibits were filled with historical objects and mementoes testifying both to their past achievements and the rapidity of contemporary progress visible in a comparison with that past. The signature pavilions of foreign nations often reproduced famous national historic buildings, castles, and royal palaces. Europeans contributed thousands of objects to the central industrial and consumer halls of the great core of the Fair. Their art works filled the collections at the Palace of Fine Arts. Colonial people, however, were labeled by anthropology according to types; they were snagged in the darkness of prehistory. Their manufactures were understood as handicraft; their art was sequestered in aboriginal cultural displays; their religion described as primitive or pre-Christian; and they were subjected to testing in the anthropometry and psychometrics laboratories in the Anthropology Building to demonstrate their mental deficiencies. Even the most positive depiction of Filipino painting, for example, suggested that artists from the colony exhibited an "Oriental mode of feeling . . . with certain servitude which favors the preservation of many of the details of the original."[46]

The context for this must be considered in terms of the discussion mounted after the acquisition of the Philippines. Political arguments about the possibility of citizenship for Filipinos became entangled in questions of American racial attitudes. How could one determine the races of various ethnic groups in the archipelago? Would they all become American citizens? McGee's effort to depict Filipinos marching along the path toward civilization was merely one side of a larger discussion among scientists, anthropologists, and amateurs about the future of the races of humanity. But, while there was considerable discussion and debate among these groups about where to place various Filipino groups, American soldiers, who fought and suppressed the Philippine independence movement and who made a noteworthy appearance in St. Louis, had no hesitation: they were "niggers."[47]

This identification of Filipinos with American blacks was a particularly potent, even dangerous, designation in this period of intense and virulent hatred directed at former slaves, and it directly influences our examination of the Jessie Tarbox Beals photograph. If we recall the enormously popular works of Thomas Dixon (*The Leopard's Spots*, 1901, and *The Clansman*, 1904), we can understand the potential gravity of Mrs. Wilkins's transgression. These immensely popular novels sold millions of copies in this era, with *The Clansman* becoming a popular stage play before its later reincarnation as the basis for *Birth of a Nation*.[48] Dixon situated his novels in the tense days of White Reconstruction, when the Clan organized itself to rescue white men from the political corruption of carpet-bag governments and white women from the clutches of former slaves.

Dixon's novels were curiously constructed, containing large numbers of stereotypic characters as well as several impassioned speeches about freedom and equality (that turn out to be deceptive and misguided). He cleverly pictured Lincoln as a hero who hated slavery and slaves. But most significant, he described, in horrifying detail, his personal physical revulsion to blacks. His language makes disgust the primal motive of his narrative. As one character in *The Leopard's Spots* declaims about white civilization, "This racial instinct is the ordinance of our life. Lose it and we have no future. One drop of Negro blood makes a Negro. It kinks the hair, flattens the nose, thickens the lip, puts out the light of intellect, and lights the fires of brutal passion." Frequent, similar passages of physical repugnance fill both novels with numbing repetition. Indeed, the descriptions of black characters could be lifted, replaced, or mixed without changing the meaning, for every Negro was, in Dixon's characterization, a latent jungle beast, vile to behold, foul to smell, crude, lascivious, misshapen, and repulsive in manner. As he put it in *The Clansman*, "for a thick-lipped, flat-nosed, spindle-shanked negro, exuding his nauseating animal odour, to shout in derision over the hearths and homes of white men and women is an atrocity too monstrous for belief."[49] This racism of the senses overwhelms any appeal to justice or humanity.

Dixon's aesthetic and sensuous outrage, his horror at the touch and sight of black Americans, denied them citizenship and soul, transforming them into animal-like creatures, threatening civilization itself. This mood of virulent racism had inspired W. E. B. Du Bois's collection of essays *The Souls of Black Folk*, mentioned earlier. Du Bois wrote of black struggles, black achievements, and black culture in terms that directly confronted the custom of segregation and the violence and hatreds that lynching betrayed. One of the most poignant of all these essays focused on the "Sorrow Songs," the folk spirituals and hymns written to raise the human spirit above the wretched horizons of slavery and exploitation. In this essay, and through his kindred appreciation of elite refinement, Du Bois suggested that culture itself might be a common ground for reconciling the races.[50] Mrs. Wilkins's cakewalk gives us a glimpse of just these possibilities of interracial communication through a shared popular culture.

If Mrs. Wilkins danced across dangerous cultural boundaries with the (almost) naked Igorot boy, with her familiar touch of a dark person, in her acquiescence in allowing an Igorot to assume the male position of the lead, she broke through the taboos of American segregation and racial etiquette and entered the more complex world of what Edward Said has identified as "Orientalism." This designation, beyond its connotations of subservience and exotic culture, was also, as the author notes, a place to find "sexual experience unobtainable in Europe." What attracted writers (and tourists) to oriental civilization at the turn of the century sometimes suggested "a different type of sexuality, perhaps more libertine and less guilt-ridden." Even this, Said remarked, could become commoditized and standardized and made available in mass culture so that readers did not have to actually displace themselves physically, but only imaginatively, to the Orient.[51] This formulation is a particularly apt functional description of the Street in Cairo and other Near East pavilions and concessions that dotted the midways of world's fairs throughout these decades—exoticism brought home to Chicago and St. Louis. But, rather than ogling Egyptian dancing girls on the Pike, Mrs. Wilkins was performing in an alien setting of bamboo huts and spear-wielding hunters, before an audience of the photographer, other visitors to the Fair, and, of course, the Filipinos themselves.

Joseph Conrad's widely read and brilliant novella *Heart of Darkness*, published in 1902, warned of the consequences of this interaction. A bitter condemnation of Belgian colonial policies, the story replied to Rudyard Kipling's poetic jingle about taking up "The White Man's Burden" with an ironic and cynical illustration of relations between white men and natives of the Congo. The story of Mr. Kurtz, the colonial agent hired to buy ivory and send it down the river and thence to Europe, is, in fact, a parable of the fatal intercourse of imperial decadence and savagery. Kurtz plundered the wilderness of its ivory ("the vilest scramble for loot"), and, in turn, the wilderness robbed him of civilization, making him the mirror image of the sensual and cruel monster that was, in fact, the European stereotype of the African.[52]

Conrad's tale could be, and was, read at the time as a warning against imperial adventures by the United States. Novelist Elia W. Peattie wrote a column making exactly this point for the *Chicago Tribune* in the spring of 1903. Conrad's purpose, she wrote, was "to expose the White man's devilish assumption—his relentless and evil usurpation of rights, and his ultimate curse." The point of Conrad's fictional confrontation between "the evil power of the savage, and the colossal assumption of tyranny," she continued, was to argue how the intercourse married the worst of both worlds. "Whatever is worst in each," she concluded, "comes to the surface, as it has probably, in the saddened Philippines and in many another land where men have been coerced into relations with the commercial, individualistic world."[53] This dark side of Orientalism exposed a fiendish temptation, according to Conrad, of power, primitivism, sensuality, and degeneracy. His somber warning was another element in the thick context that informed Mrs. Wilkins's lighthearted dance.

But what of the dance itself? What did it mean for Mrs. Wilkins to teach an Igorot to do the cakewalk? What significance did this peculiar dance hold in American culture at that moment? These are complex questions, compounded further by the near nudity of her Filipino partner. What made it safe for her to grip the hand of this native when in other circumstances such intimacy would have been extraordinarily controversial and risky? Returning to the photograph, we may be able to see why. Part of the answer may well lie in Mrs. Wilkins's stylish outfit. Wearing a long white dress and a hat capped with a peacock feather, perched atop her fashionable pompadour hair style, with her body shaped into the familiar and popular S-curve by modish undergarments, Mrs. Wilkins overwhelms the photograph with her size, her power, her whiteness, her physical presence, thus greatly diminishing the near-naked Filipino. Everywhere else on the fairgrounds there were similar vivid contrasts between Western dress and native "costume," between the rich and varied gowns on display in the French exhibit at the Manufactures Building and on the Pike's Palais du Costume and the "characteristic" and seemingly unchanging garb that made up part of the symbolic identity of Native Americans, exotic Middle Easterners, and Filipinos. The contrast in this photograph is one repeated frequently, perhaps almost unconsciously, in most of the photography showing the audience mingling with exhibited peoples. Perhaps Mrs. Wilkins could be confident because fashion was a symbol of Western superiority, a sign of her respectability.

When Mrs. Wilkins offered instruction in the cakewalk, what sort of dance was she teaching? The cakewalk has a fascinating history, born out of satire, developed as a slave spectacle, transformed through black-face minstrelsy, and, by 1904, emerging as a popular dance among both black and white Americans. In fact, it was one of the first of many future crossover dances and musical forms (such as ragtime) that originated in the black cultural world and, by evolution and masking, mutated into something that whites could also perform as if it were their own.

It is difficult to locate the precise origins of the dance, although, apparently, it began when slaves observed and lampooned the formal dances of white plantation owners: the parade, the minuet, the quadrille. With elaborate bowing, tipping of hats, and high-kicking promenades, the cakewalk appeared to white southerners, on the other hand, as a clumsy attempt of blacks to imitate their betters. It is also possible that the exaggerated strut that characterized the dance and its satiric purposes may have reflected elements of African dance brought with the slaves to America. As Eric Sundquist has written, the dance "fused its African roots with a satiric commentary on big house fashion."[54] Through their misunderstanding, white plantation owners transformed the dance into a special performance for themselves, often at harvest time,

FIGURE 13. Poster advertising "Uncle Tom's Cabin," a cakewalk entertainment that played successfully in New York around the turn of the century (1898). In this production, the actors are referred to as "our colored Exclusives." (Library of Congress Prints and Photographs.)

when the mistress of the plantation would award a prize cake to the dancing couple judged the best performers. By the end of the nineteenth century, the cakewalk had become a staple of minstrel shows, performed by both blacks and whites in blackface. Its immense popularity and appeal to all audiences transformed it into a national rage, spawning ballroom and specialty dance contests at Coney Island, for example. At the same time, its origins in African and slave sources and its satiric purposes became obscured. By 1904, even John Phillip Sousa performed a number of cakewalks at the St. Louis World's Fair, although in a much diluted and less raucous format than what appeared in Broadway or Harlem theaters.[55]

The most popular cakewalk song of the day was probably "Under the Bamboo Tree," performed first by the Irish singer Marie Cahill in the blackface Broadway musical "Sally in Our Alley." Written by black composers James Weldon Johnson, J. Rosamond Johnson, and Bob Cole, the song sold over four hundred thousand sheet music copies, a good indication of its wide circulation.[56] Perhaps Mrs. Wilkins was singing these lyrics when she taught her Igorot friend to dance.[57]

FIGURE 14. Contemporary white version of the cakewalk. This poster, created in 1898 by Allied Printing of Chicago, shows a white couple dancing this popular dance. (Library of Congress Prints and Photographs.)

UNDER THE BAMBOO TREE

Verse 1
Down in the jungles lived a maid,
Of royal blood though dusky shade,
A marked impression once she made,
Upon a Zulu from Matabooloo;
And ev'ry morning he would be
Down under the bamboo tree,
Awaiting there his love to see
And then to her he'd sing:

Chorus
If you lak-a-me lak I lak-a-you
And we lak-a-both the same,

> I lak-a-say,
> This very day,
> I lak-a change your name;
> 'Cause I love-a-you and love-a you true
> And if you-a love-a me.
> One live as two, two live as one,
> Under the bamboo tree.

Was Mrs. Wilkins thereby teaching her young friend to be black? Was she unconscious of the black origins of the song and the dance? Was she teaching him to be an American? Or was she purposely crossing racial lines, knowing full well what her actions symbolized? We cannot know, but we must consider all of these possibilities.

Our interpretation of the dance, however, cannot terminate in 1904 or with Mrs. Wilkins or even the Igorots. The cakewalk and the Fair also became linked in the important transcriptions of memories of this era. The young and future poet T. S. Eliot visited the Louisiana Purchase Exposition over the summer of 1904, where he might have seen the cakewalk, other dances, and "coon songs" presented that year during "Louisiana, A Spectacular Extravaganza in a Prologue and Two Scenes."[58] Or perhaps he knew the song from local performances in St. Louis. In 1924 Eliot composed sketches for a play entitled "Sweeney Agonistes." Performed first in 1933, and only infrequently thereafter, the piece quoted the lyrics of "Under the Bamboo Tree" in a passage that suggested Eliot's later quotation from Conrad's *Heart of Darkness* in his great poem "The Wasteland." Both the lyrics and Mr. Kurtz's last words were linked in his mind to the confrontation of high culture with primitivism, jazz, and minstrelsy (in other words, popular culture). As critic Tatsushi Narita has written, "The encounter of the young Eliot with the Philippines is perhaps one of the most striking incidents of his brief life in St. Louis."[59]

The best-known and pertinent reprise of the song came as a musical number in *Meet Me in St. Louis*. Sung by Judy Garland, as she dances with her film sister, Margaret O'Brien, the song had by then lost most of its racial codes and associations, which are vaguely present at best. The two perform the cakewalk without its syncopated cadences, its outrageous scamper, its humor, and its historical significance. Danced by the pair on a crowded living-room set, filled with family members and their beaux as audience, it has mutated into a modern, derivative Broadway number complete with straw hats and canes, a bland evolution from the strut that Mrs. Wilkins taught.

In fact, this performance of the song demonstrates one way Hollywood films empty the past of its racial and ethnic identity, in this case transforming the past into the setting for a happy white, middle-class family memory.

FIGURE 15. Judy Garland does the "cakewalk." Judy Garland and Margaret O'Brien (as "Tootie") perform the cakewalk and sing "Under the Bamboo Tree" in *Meet Me in St. Louis*. Their dance is more like a step borrowed from a Broadway musical than the cakewalk popular in 1904. (MGM, 1944.)

The African American origins of the cakewalk have disappeared save for the puzzling lyrics. Indeed, the presence of the dance in the film itself is jarring, explainable only as a historic quotation designed to give verisimilitude to this period piece. It exists as an artifact offered up to the 1944 audience and to us: an empty sign, except to those familiar with the song and its history.

The final context in which to place Mrs. Wilkins and her dance is perhaps the most obvious and controversial one. There is no denying the sexual connotations of this encounter in 1904, even if her intentions were completely innocent. In fact, however, relations between Filipino men on the fairgrounds, especially members of the Filipino Scouts and marching band members, and white women attending the Fair or living in St. Louis had exploded in mob violence. Beyond smirks over the nudity of some Philippine village peoples, there were a number of real encounters that led to small riots and, on occasion, beatings of Filipino men, touching off larger public anxieties about interracial fraternization.[60]

Dating Filipino men was controversial in both the black and white St. Louis communities, although far more so among whites.[61] On July 24, the

Chicago Tribune reported a brawl between white men and Filipinos at the Café Luzon on the Philippine Reservation. In the end, "six white women* and a large number of [Filipino] Scouts and Constabulary were arrested," the paper noted. Although the ostensible cause was the desire of the Filipinos to prevent the restaurant from closing for the evening, the incident has all the earmarks of an ethnic and racial fight over white women. This followed a pitched battle on the Pike a week earlier when a squad of American Marines attacked Filipino Scouts escorting white women near the Mysterious Asia concession. A series of running skirmishes ensued as the Marines expressed their outrage at the "smoked brethren" and their white dates. Ultimately, a mob of two hundred men from both sides ended in a "struggling heap," as the *Des Moines Daily Capital* put it.

In the course of these events, white American men refused to recognize any subtle distinctions of ethnicity and anthropology. What they could not see physically was nonetheless visible culturally. U.S. Marines and the Jefferson Guards, for example, assaulted Filipino Scouts escorting white female schoolteachers on the grounds, threatening a lynching. As the *Des Moines Daily Capital* explained, "They have declared war upon the coffee colored brother and the cream colored sister when seen together."[62] This context makes Mrs. Wilkins's dance even more controversial, for not only was she stepping high to the cadence of the cakewalk, she was stepping over and transgressing an absolute color line, symbolically enacting in her dance the very offense that elsewhere caused violence to break out. Perhaps we should also ask if she was one of those schoolteachers?

With her dance, Mrs. Wilkins entered a site of fraternization that historian Kevin Mumford, writing about Chicago and New York vice areas during these early years of the twentieth century, has called "interzones." These were areas of cities where whites and blacks could meet for social and sexual intercourse. Limited in geography, generally hidden, and carefully contained, these areas, nonetheless, provided a place to meet across racial and social boundaries even if they maintained the rules of racial hierarchy and gender exploitation. If we extend this concept and enlarge it beyond the primarily sexual connotations that Mumford describes, as something more akin to the traditions of slumming that historians have found prevalent in the late nineteenth century, we can see that in some respects, the Fair offered the possibility of a vast zone of suspended rules and social codes, where persons of different races could and sometimes did interact.[63] Given the striking absence of black Americans from the audience, it may have been possible to imagine or experiment briefly

---

*Every account I have read stresses the race of the women.

with social interactions between Caucasians and dark ethnic groups without the constant reminder of the harsh rules about race prevailing outside the gates. The Fair, despite the presence of the Jefferson Guards to enforce good behavior, and despite the heavily didactic messages of the administration, was treated as a vast amusement park, where the audience could turn the concessions and displays to its own purposes.

Regardless of limitations imposed by formality in dress and manners and the prevailing context of racial and ethnic etiquettes operative elsewhere in American society, the Fair offered a possibility to break some of those rules in somewhat the same way that unregulated, new urban spaces did. This had never been McGee's intention; in his mind, anthropology was not just an academic discipline but a way of disciplining the imagination so that it would recognize separation, hierarchy, and progress along a continuum. Nor, certainly, was it the purpose of the giant Philippine Reservation and its U.S. government sponsors to encourage fraternization. One of the temptations of imperialism that Conrad had so vividly sketched was the (dangerous) possibility of enacting intimate social interchange with primitive peoples, of going native and then never coming back. Distant colonies were governed, at best, by an extension of home rules, loose and ill-fitting in most cases, but intended to contain and alienate the different customs of another culture. And interaction was to be closely guarded.

"Savagery" came to St. Louis, to be gawked at, of course, and to be studied, but for some there were more adventuresome possibilities. Just as American men were attracted to the risqué dancing on the Pike and the seminudity of Philippine women, so American women were drawn to the scantily clad Igorot men. This represented much more than a gaze across boundaries, just as the cakewalk was more than a popular dance. Contained in them were possibilities for new human interaction. Such hesitant encounters developed throughout the rest of the century in other settings as Americans confronted each other and the rest of the world. By using a popular dance to jump across divisions of race, privilege, and civilization, Mrs. Wilkins demonstrated the usefulness of popular culture to translate (for good or ill) the abrupt juxtapositions and disruptions of modernity into something comprehensible and less threatening. Like midways elsewhere, and urban environments in general, the Pike demonstrated the modern tendency to blur the distinction between entertainment and information. Especially on the Pike, but true elsewhere at the Exposition, interactions with all that was strange and new in history, foreign peoples, empire, religion, science, ethnicity, technology, sociology, and anthropology seemed best understood and appreciated when reformulated into spectacles of the imagination, a new nation conceived through pleasure.

CHAPTER SIX

THE BEHOLDER'S EYE
Making Experience

Despite the best efforts of planners and public relations experts to control and guide visions of the Fair, and regardless of our implicit assumption, as historians, that they were largely successful, ordinary, individual experience keeps breaking through. I have been trying to show that memory, for all of its obvious failings, suggests a very different sort of world's fair encounter, and I have discussed how different genres of photography reveal fundamental divergences in depicting that event. Mrs. Wilkins's dance is only one vivid example of what we can learn by carefully examining a photograph within and beyond its immediate frame of meaning and how larger cultural assumptions engaged the Fair. But can we gather all of these sources together to develop new ideas about the significance of the Fair in St. Louis? Can we look through them to the (elusive) experience of visitors? And will this analysis tell us anything about how generally to approach such events?

There is, of course, a serious problem with this approach. We cannot reconstruct actual experience either for individuals or groups. It is impossible to delve into minds in the past and unlikely, anyway, that we could find a single moment in the kaleidoscope of shifting impressions that would reveal a stable meaning: *the* meaning. And yet both collective memory and history insist that we can. That the meaning of an event is known to its participants is implicit in collective memory and historical accounts. There is an assumed coherence between our interpretations and actions in the past even if this assumption is rarely explored.

And a further word of caution: we should not dismiss as frivolous the argument that the participants might have missed the historical significance of their own actions. The meaning we ascribe to the past is always future-oriented; it evolves because some implications become stronger as time passes, and new ones emerge. That is why historiography thrives, to take account of these changing perspectives. But I still believe we can develop new ideas about the meaning of such grand public events at the turn of the nineteenth century, new ideas that put the participant back at the heart of the experience. These conjectures depend upon our paying much closer attention to what visitors thought they were doing, what they said at the time and afterward about it, and how they voted with their feet, their money, and their attention. This demands scrutiny of two different types of documents: individual accounts in various forms and the evidence of group behavior suggested by concession reports and other measures, including visual materials. This chapter will explore both sorts of records, for they are mutual indicators of experience. And like all historical source material, they achieve significance only through interpretation, imaginative reconstruction, and contextual reading. As Clifford Geertz has written, "Behavior must be attended to, and with some exactness, because it is through the flow of behavior—or, more precisely, social action—that cultural forms find articulation."[1]

Much can be learned from memoirs, oral histories, diaries, interviews, and other forms of individual reporting about the Fair. Seemingly mundane, this material reveals the importance of the ordinary. The journal of Mary McKittrick Markham, wife of the chief of the Department of Music, suggests how the life of elite St. Louis spilled out onto the fairgrounds, consuming the summer in a whirlwind of local social events. Scores of times Mrs. Markham's destinations in Forest Park were concerts, lectures, dinners, and receptions for visiting dignitaries. The visit of Alice Roosevelt (the president's daughter) was a particular high point, but she also met ambassadors, politicians, and important intellectuals. Between these important events, Markham also managed to visit a variety of displays and concessions, but, quite clearly, the responsibilities of social life drew her to the Fair. It is fitting that among the mementos she preserved—the invitations to receptions and lecture programs—was a newspaper article entitled "The Fair as a Resort." It certainly was for her social group. She and the rest of the local elite remaining in St. Louis for the summer performed the vital function of hosting visiting celebrities. They comprised the select audience to fill reception halls and concerts; they were a class on call that the Fair administration could depend upon to attend events that needed a polite ambience. In this manner St. Louis society, the officers of the Louisiana Purchase Exposition Company, their wives and friends made the Fair the meeting ground of a national and international class of intellectuals and politicians. They experienced the Fair as a group that reached out to include ambassadors, businessmen, and other dignitaries; what a summer it was for them! This was a group that transformed the Fair into a national extension of itself.[2]

At the other end of the social scale, Laura Merritt, from Cedar Rapids, Iowa, related an entirely different sort of visit, although it too was defined by her identity and interests. Arriving by train in September, she and her family went immediately to the Iowa State Building on the fairgrounds. Fortunately they met two neighbors from Cedar Rapids who suggested a reasonable boardinghouse, where they moved for their stay. Laura and her family spent the first two days watching Olympic contests and then the next two at the United States Government Building. Thereafter, their itinerary continued through most of the main display palaces, although she was clearly most impressed by the Agriculture Building, "voted" the "prettiest and best building," she noted, by foreigners and a large share of Americans. Merritt also spiced her account with references to "Chinks" and "Japs," even though the latter impressed her as "the wonder of the age." She was delighted to find that the Philippine band unexpectedly could play American patriotic songs.[3]

Merritt's diary is noteworthy in two respects. It reveals the importance of affirming a state and local identity for visitors as part of their interaction with the Fair, and it underscores the crucial part played by the state pavilions on the grounds. The family often began their day at the Iowa Building. Laura certainly reacted to the international population on the grounds and the presence of thousands of foreigners. Her references were peppered with stereotypical racial slurs, but, at the same time, her admiration for the Japanese breaks through this layer of offhand prejudice. Like most of the other diaries, journals, and other personal accounts, she does not indicate whether or not she either understood or remembered McGee's anthropological message. In this she seems typical.

I have already explored the fascinating diary and memoir of Edward Schneiderhahn. As earlier noted, he transformed his disorganized and highly impressionistic diary record into a memoir that more or less followed a guidebook format, even repeating the propaganda about the benefits of American presence in the Philippines. We cannot know precisely why he made this transformation. Nor can we tell if the revised account actually includes only his experience or something more that was added on. But there is the impression of authenticity in the initial record in its raw and unstructured form, before he revised it. The breathless accounting and the jumbled quality to this narrative, with short visits to a variety of places, the very density and multiplicity of what he managed to see, suggests, ironically, how much he did not see. This emphatically reminds us that most visitors could not have time, nor did they have the inclination, for a truly orderly or comprehensive visit.[4] Indeed, our understanding of experience must be premised on recognizing the limits to what could be seen in a day or two, or even a week. Other tourists, like Schneiderhahn, were also driven by interest and curiosity, finances (he clearly spent far more on the grounds than the average tourist), and, in his case, ethnic origin, as he frequently attended German events. Just as Laura Merritt and her family were drawn to the Iowa pavilion, so he was attracted to exhibits and entertainments that expressed his heritage.

Other diaries reinforce the conclusion that visitors saw the Exposition by following their own idiosyncratic agendas. Edmond Philibert, a carpenter, described twenty-eight visits pretty much in the order he saw them; there is little logic to his progress through the Exposition except for his tendency to divide his time each day between exhibit palaces and the Pike. One visit on July 30, for example, included observation of the Railroad and Transportation Day Parade, during which he watched the Jefferson Guards, automobiles, Boer War veterans, Igorots, and Eskimos pass by. The parade presented a sample of exhibits, highlighting both the palaces and the Pike. Its procession of people and floats jumbled the orderly logic of a "World's Encyclopedia." But the purpose in this case was not knowledge but advertising the highlights of the most popular exhibits on the grounds. Expense made this visitor a judicious consumer. Philibert spent most of his funds on admissions, food, and the internal railroad conveyance, but visited relatively few concessions, even complaining at one point about the high prices charged inside the Tyrolean Alps.

Philibert often went to the Fair in the company of his sister. Like other visitors, he frequented state exhibits and on one occasion used the Louisiana Building to meet relatives from that state. Several times he visited popular religious exhibits, such as the Creation concession and the display mounted by the Vatican. Like other writers, when he did comment on encounters with "primitive" peoples, he fell into prevailing racial attitudes and slang. He found the Patagonians, for example, to be "lazy looking." And he noted about the re-creation of a southern plantation: "It consisted of about twenty darkies singing, dancing, cake walking."[5]

On the other hand, Sam Hyde, a bookkeeper from Belleville, Illinois, recorded very different impressions. He attended a variety of exhibits but seemed offended by the chaos and racial diversity of the Pike. Commenting on these midway types, he wrote, "I was taking in these free shows [during the end of the Fair] and studying the faces from a scientific point of view. Repulsively vile, most of them. I thought it would take no prophet to detect

'The Mark of the Beast' on their foreheads." Hyde also had an aversion to the sexual innuendos of the concession barkers. Although he skipped the Igorot Village, he suggestively noted that "they seemed to have a tremendous attraction for the ladies." If anything, Hyde's account is noteworthy for his scurrilous remarks about race in general. He fit his encounters with Native Americans, Japanese, and other groups into categories and terms of his preexisting prejudice rather than the comprehensive and relatively nuanced anthropological schemas promoted by the Fair anthropologists.[6]

Oral histories, reproduced in printed recollections and later in the testimonies taken in 1979 to commemorate the seventy-fifth anniversary of the Fair, reveal much the same unpredictability. In 1954, fifty years after the Fair, the *St. Louis Post-Dispatch* printed a number of reminiscences. Theodore P. Wagner, for example, recalled that he and a houseguest went to the Igorot Village and asked a Filipino if the cooking pot he was stirring contained dog. The disappointing answer was no. His other memories were of Chief Geronimo (also a letdown) and various other concessions and displays: the Boer War, the Transportation Building, and so forth. Wagner's account, as is often the case in twice-told memories, has a polished, narrative gloss. This is how he related the experience (in adult language) of being an eight-year-old boy at the center of multiple comings and goings of an extended family: "The Fair attracted relatives and friends like a molasses jug of that distant day attracted flies. They came in doubles and triples and bunches until our home on South Jefferson, opposite the old Concordia Seminary, fairly crawled with visitors."[7]

Wagner's account reveals a young boy's swagger and a practiced narrative style. We should not be put off by this, though, because he also reinforces the impression gained from other oral histories and accounts that the Fair was very much a family event. As a thriving midwestern city with a large and burgeoning population, St. Louis welcomed immigrants of all sorts: Germans and Irish from Europe and especially young men and women from small towns and farms all over Missouri and surrounding states, most of them recently arrived. As Emma Helena Kuhn Mohr recalled in her oral history interview in 1979, she traveled from Trenton, Illinois, to stay with her sister and her family living in St. Louis at the time.[8] Countless other visitors to the Fair had relatives in the city, and their visits doubled as informal family reunions. This apparently mundane fact has far more expansive implications for our understanding of the Exposition, because it reproduced on a small scale the scores of large, national family reunions held on the fairgrounds over the summer.

Oral histories taken for the seventy-fifth anniversary of the Fair in 1979 were confined to a group of elderly residents of the city who had been children in 1904. Their youthful angle of vision may well be responsible for the lack of narrative perspective that marks many of them. As we know, memory itself sometimes operates to recontextualize events or sometimes to place them in jumbled disorder or mix them with other accounts derived from secondhand experience, reading, and conversation. What an individual actually experienced could be enhanced by what others told him or her or by common knowledge or rumor. Having the credentials of a witness might also be a license to exaggerate and embroider.

While the unreliability of these accounts can be frustrating for oral historians—as it was, for example, in the case noted earlier of the interview with Willard B. Shelp by two representatives of the Missouri Historical Society—even the information gathered from him is revealing. A twelve-year-old boy in the summer of 1904, Shelp visited the Fair numerous times. Because his parents had a large house, they decided to rent it to an "official of the Fair from a foreign country" while the family moved across the street to the Lindell Plaza Hotel. Shelp's discussion of the Pike and its entertainments—the Ferris wheel, the Eskimo Village, dancing girls (whom he did not see), the Battle of Santiago, Mysterious Asia, as well as the Boer War reenactment, ice cream cones, the Igorots, the Cascades, and electric lighting at night—constitute a list that almost entirely excludes anything from the grand display palaces that made up the essential core of the Fair. For Shelp, the Fair was a gigantic midway, filled with delights and oddities, a place of wonder rather than an opportunity for controlled and orderly learning.

Like the narrators of many of the other oral histories taken at this time, Shelp places himself as a witness at the center of the most important event in the history of the city. "It was a fabulous time that I was privileged to live and be alive today," he concluded.[9] His large sampler of sights and impressions gives the sense that he went to the Fair scores of times. Despite the boasting tone of his account, we can learn much from what he has to say. First, we see another crucial role played by upper-class St. Louis residents during the Fair. As a wealthy family, his parents joined others, particularly those who lived in Pershing Place, who rented their homes over the summer, making that section of town a quarter for visiting dignitaries. Second, Shelp suggests that the city itself became a temporary theatrical capital of the country because "many of the famous theatrical stars, men and women, moved to St. Louis and stayed here during the summer." Much the same sort of thing could be said about musical performances, with important figures like John Philip Sousa and Scott Joplin playing in St. Louis, as well as international military bands and classical performers. Based in part on this immense cultural influx, the huge number of tourists, hotel construction, and ancillary economic activity associated with the Fair, the city experienced a small economic boom similar to the impact on Chicago during 1893, when the Columbian Exposition temporarily staved off the local effects of a severe national depression.

Perhaps the most revealing memory source we have is the survey prepared for the Missouri Historical Society in 1994, prior to the 1997 exhibit "Meet Me at the Fair," mentioned earlier. The survey sampled all areas of the city and divided respondents by age, race and ethnicity. Its findings indicate the strength of contemporary collective memory of the Fair, but also much more; they suggest that memory is an ethnic and gendered matter. As I have already discussed, almost 50 percent of the respondents (the highest total) chose the Fair as one of the three most important events in the history of the city (beating out the Gateway Arch, baseball, the flood of 1993, and Lindbergh's transatlantic flight). Breaking down these figures by the race or ethnicity of the respondent, however, the survey revealed a profound difference in the answers. Even with three possible answers, blacks in St. Louis scored the lowest in naming the Fair as a memorable event, with only 30 percent mentioning the Louisiana Purchase Exposition at all. On the other hand, figures were much higher among whites, with 50 percent. The highest tally was found to be among "whites with a particularly strong ethnic identification, such as German or Irish." Almost 70 percent of these two groups cited the Fair as the most significant historical event of the city's history. Measuring where participants first learned about the Fair, by far the highest percentage was in the family, followed by school, the media, and, finally, the film *Meet Me in St. Louis*. When asked why the Fair had such great importance for the city, the highest percentage cited "prestige," followed by "historic event" (which might be construed to mean the same as prestige).[10]

In a great many cases, memories and oral histories of the Fair begin with family stories. The family is also generally cited as the place where the respondent first learned of the event. Also fascinating is the way personal memory has become entwined with the prestige of St. Louis itself, constituting an apparent high point in the history of a city whose modern fortunes have sometimes seemed less positive. Of greatest consequence is the ethnic and racial component in recollections of the Fair, for this can tell us both about contemporary collective memory as well as something about the significance of the Exposition as it was experienced in 1904. That the black community of St. Louis does not recall the Fair as a grand moment of St. Louis history suggests that it was not a great time in their history. The affronts of segregation on the fairgrounds, the cancellation of Negro Day, and other suffered indignities, though not remembered directly, probably still serve to dampen enthusiasm for what were decidedly not happy or nostalgia-etched times.

But why this heightened memory among St. Louis immigrant groups, particularly the Irish and Germans? Does this memory have anything to do with the particular role of these two populations in St. Louis in 1904 or their representation and activities on the fairgrounds? What special appeal did the Exposition have for these ethnic groups? One answer may be suggested by a closer look at what visitors actually did when they went to the Fair, what concessions and displays they entered, and what might have drawn them to these places.

I have already noted that the unequal proceeds from the Philippine Reservation underscore the huge attraction of the seminude and "barbarous" Igorots. In fact, the least visited village, the Visayans, purportedly were a people who represented the highest form of civilization on display and were most susceptible to American influence. Even if visitors understood McGee's broad scheme of race division and evolution, they chose not to experience it by passing from one village to the next up the hierarchy of progress. Instead, their attention fixed on the more exotic exhibits of "primitive" peoples. I have also shown that, while race was a constant issue on the fairgrounds, the anthropological justifications of colonialism promoted by the United States Philippine Commission were often translated by visitors into their own existing racial and ethnic attitudes and probably not reshaped or changed much by these exhibits. This conclusion is underscored by a report of the Jefferson Guard. Explaining the belligerent behavior of ex-American soldiers toward Filipino men escorting white women on the grounds, the report began: "The Philippine Scouts and Constabulary, while not negroes, have a dark skin, and were classed as negroes by most of the people visiting the Fair." Unfortunately, the Jefferson Guard itself took part in some of these racially charged skirmishes, the most serious being a violent clash near the Ferris wheel including twenty or thirty Filipinos and a large number of artillery men and Marines, who exchanged gunfire and rocks. At other times, the Guard clashed with Filipinos who had armed themselves. But their motivations seemed to stem directly from the troubled race relations that divided contemporary Americans. Even the organization of the Filipino soldiers reflected this American racial experience. The troop consisted of 420 men with 12 white American officers. This racialized command structure reflected the organization of black troops in the U.S. Army during and after the Civil War.[11]

Given this context, what do the proceeds from concessions on the Pike reveal about race and ethnicity? What do they tell us about what visitors were actually willing to pay to see? To answer such questions, we must first recognize the meaning of the Pike as a cultural construction. Despite their commercial purposes and the fantastic shapes and designs, concessions on this midway frequently represented popular-culture versions of the social and cultural concerns displayed elsewhere at the Exposition. They presented popularized forms of ideas about race and ethnic issues, sexuality, music, religion, history, anthropology, technology, science, tourism, and warfare, all of which were subjects to be found in the main areas of the fairgrounds. No doubt this double duty, combining pleasure with a reference to significant issues, increased the interest of the huge audiences that strolled up and down the Pike, watching barkers, sideshows that spilled into the street, parades, and, of course, the polyglot crowd itself. The Pike was a popular-culture incarnation of the more abstract ideas and processes displayed in the palaces, in terms that patrons could readily understand and relate to. Although we cannot be absolutely certain about the breakdown of concession proceeds (how much was spent on entrance fees, then on souvenirs and food once inside), there are tentative conclusions we can draw about how many patrons attended the various paying concessions. Comparing these proceeds to those of the Philippine Villages makes them even more significant.[12]

In terms of gross proceeds, the Tyrolean Alps was by far the most successful, earning over $1,000,000. The next highest was the Boer War, which earned over $600,000, followed closely by the Irish Village, then Creation, and the Ferris wheel. Then came Hereafter, another religious entertainment, followed by a firefighting display, a scenic railway ride, and a re-creation of the Galveston flood. Because entrance fees were $.25 per adult and $.15 for children, we can conservatively estimate (even subtracting for the expense of internal attractions) that there were probably two million visits to the Tyrolean Alps, perhaps a million to the Irish Village, and then fewer amounts proportionately to the other major concessions.[13] This compares to our estimate of one million total visits inside at least one village located inside the Philippine Reservation. Taken separately or together, concessions on the Pike constituted a huge draw, much of it concentrated on displaying exotic peoples in an entertainment format. This provides a strong argument that its versions of race and ethnicity received more popular attention than the formal anthropology of McGee. And even then, many patrons drawn to the Philippine Reservation probably had the same motivations that enticed them into paying concessions on the Pike: heightened curiosity about the erotic and exotic.

Perhaps we can credit the extraordinarily high attendance at the Tyrolean Alps and the Irish Village exhibit to the large number of Germans and Irish living in St. Louis at the time and in nearby areas of the Midwest like Chicago. I have shown, in exploring oral histories and other personal accounts, that residents of St. Louis attended the Fair frequently. And just as visitors

FIGURE 16. "Two Men Walking on a Road through the Tyrolean Alps." This photograph suggests the comprehensive visual fantasy of the Tyrolean Alps exhibit. (Jessie Tarbox Beals Collection, Schlesinger Library, Radcliffe Institute, Harvard University.)

from Iowa registered at the Iowa pavilion, met friends, and admired displays from their state, it is entirely plausible to suppose that first-generation immigrants or persons of German and Irish descent found the same sort of identity affirmation in German and Irish places on the fairgrounds, not just on the Pike, but also in exhibits elsewhere. These existed, like many places on the fairgrounds, as "sites of authentication," places where the multiple identities of Americans could be visualized and asserted. These magical places offered such groups the possibility to maintain a sense of continuity with family origins and historic places as well as a way to affirm their identity as immigrants belonging to the nation.[14]

During this turn-of-the-century era, both Irish and German populations had begun to make considerable progress in remolding the stereotypes that had once defined their identities. Historian William Williams finds that Ireland ("The Emerald Isle") became an enchanted rural paradise in contemporary commercial popular culture, populated by handsome young men and women. This rural Eden corresponded closely to the American idyll of the country homestead. Most important, Irish popular culture, particularly in song and other performance, might help convince Anglo-Saxon Americans that this past was also their past. This was, no doubt, also the achievement of

FIGURE 17. "In the Irish Village—Pretty Lassies in Jaunting Cars before 'Cormac Chapel.'" The Irish exhibit on the Pike presented Irish stereotypes, as in this image by Underwood and Underwood. (Library of Congress Prints and Photographs.)

such exhibits as the Blarney Castle and, for Germans, the Alps displays.[15] Or as Matthew Jacobson puts it, Germans and Irish at the turn of the century "assumed the status as honorary 'old-stock' Americans."[16]

According to the published guide to the Tyrolean Alps concession, this fantasy re-creation of the homeland was designed to make it possible for German immigrants to "tighten the links between the Germans in America and their new home. On the other hand they will awaken in the minds of Americans a feeling for the beauties of German landscapes and for the honest straightforwardness of the men and women of the Alps who will occupy the comfortable huts, shops and lanes of the Tyrolese village." Thus, the Alps invited both immigrant and nonimmigrant to share the same vision of Germany as an imaginary, even mythical place, a Neuschwanstein before Disney's famous model of that castle constructed fifty years later in Anaheim, California.

The Tyrolean Alps was a joint project of local St. Louis German Americans and German and Austrian officials. The project was financed by Adolphus Busch (also on the Fair board of directors) and other St. Louis businessmen. Its purpose was to create an "intercultural force." As one historian of German culture in America put it, "the German and Tyrolean Village—the most attractive and best attended location of the Pike—is unthinkable without the German-American community in St. Louis and without the power that Adolphus Busch represented at the time."[17]

In 1900 German immigrants and their children made up 56 percent of the population of St. Louis, followed by the Irish and their children at 15 percent. Together, these two cohorts constituted about 70 percent of the city's population, exercising a huge influence on the culture of the city. St. Louis was a city of *Turnvereins* (athletic clubs), singing clubs and choirs, German-language schools and newspapers, as well as St. Patrick's Day banquets. Citizens of German origin, however, were scattered throughout the city and not concentrated in a few ghettoized wards, which may account for the importance of *kultur* in uniting this ethnic group.[18]

Besides the huge German national pavilion, displays spread out in many of the principal buildings on the grounds, and the Tyrolean Alps exhibit, there were noteworthy cultural events like the Saengerfest. This music festival, held at the Liberal Arts Hall, boasted the participation of over one hundred U.S. German singing societies. The Catholic charities Central Verein also held its national convention in St. Louis over the summer of 1904. After the Fair and up until the near-extinction of German American institutions during World War I, such public events and commemorations continued to punctuate St. Louis history, many of them featuring "the ever present German village," as one observer put it.[19]

Similarly, the Irish Village on the Pike was a center of Irish culture and activity. Celebrity singer John McCormick visited the exhibit to conduct concerts, and several plays by William Butler Yeats were performed. Local, national, and Irish ethnic commemorations were mixed together with religion on the fairgrounds when John Glennon, the archbishop of St. Louis, attended a celebration of (Catholic) "St. Louis University Day at the Fair" that included, as guests, the president of the Louisiana Purchase Exposition Company, David Francis, and several thousand others.[20] As Marion R. Casey has written, Irish Americans in this period struggled against ethnic stereotypes and slurs and, in particular, the racial designation "Celtic." But the Exposition did its small share to overcome this prejudice. As Thomas R. MacMechen, press censor for the Pike, wrote in comments entitled "Ireland at the World's Fair," "No true Celt would dream of going any place where he couldn't have fun, even if it was at his own expense. This will not be the case here, however, for he can find fun galore, and of a high and wholesome character." In the nostalgic re-creation of Blarney Castle and other symbols of Irish culture and in the displays of handmade goods and products in the Irish exhibit, we can see the dual elements of this ongoing effort to create a mythic Ireland to reflect favorably on Irish Americans.[21]

There is further illustration of the central role played by the Irish and Germans throughout this period in St. Louis history. In one celebrated instance, during the allegorical Pageant and Masque of St. Louis, performed over four days at the end of May in 1914, and attended by thousands of citizens of the city, the Irish and German presence in the city was specifically acknowledged and recognized. Other groups, aside from the revered French founders of the city, were bunched in the generic category of "immigrants." And black Americans were not represented at all.[22] This privileged ethnic identity was echoed and reproduced in recent memory culture when the 1979 Fair anniversary celebration featured both an Irish Beer and Wine Garden and a German Beer Garden at the center of a re-creation of the Pike.

If Irish and German ethnic groups could affirm one portion of their dual identities as Americans and German or Irish by visiting concessions that were designed to generate and exploit nostalgia, they also shared with other Americans a sense of who they were *not* by viewing exotic cultures on display. Here they became part of the larger white audience. This spectatorship extended to the growing fascination with "Oriental" (Middle and Far Eastern) cultures that marked the years around the turn of the century. One of the largest and best-attended exhibits of the Fair, although not located directly on the Pike, but between it and the main section of the Fair, was a scale-model, ten-acre reproduction of the city of Jerusalem. This remarkable exhibit combined both the possibility for religious-based vicarious tourism for Christian Americans as well as secular experience of encounters with its exotic Arab and Jewish denizens.[23]

Orientalism was very much in vogue during these years. Historian Melani McAlister notes that in 1903 Siegel-Cooper's New York department store ran a six-week-long "Carnival of Nations" ending in "Oriental Week," which included theatrical representations of a Turkish harem, parades of dancing girls, and a Cleopatra. There were many other signs of the popularity of the exotic East: Robert Hitchins's popular novel *Garden of Allah* (1904), for example; the hugely successful Puccini opera *Madame Butterfly*, written in 1904 and performed first in the United States in 1907; and the vivid influences on Postimpressionist art.

Orientalism was ubiquitous on the Pike, from the Street in Cairo, to

Mysterious Asia, to the Moorish Temple, to Fair Japan, Chinese Villages, Constantinople, and the Shanghai Restaurant. Of the fifty or so major concessions, this generic group amounted to almost 15 percent and the largest category of exhibits in the amusement section. This Oriental inflection was not only apparent inside the paying exhibits but also omnipresent in the street scenes recorded by photographers who captured costumed concession workers sauntering alongside patrons. Reporting on an extraordinary "Pike Parade" passing like "an opiate dream," an article in the *World's Fair Bulletin* of May 1904 reported a sizable representation from the Pike's major and minor exhibitors grouped into five clusters, each headed by a marching brass band. As the story noted, "This passing show was a harlequin smattering of religions and customs running from time immemorial down to the modern world of beauty and fashion" (alluding here to the Palais du Costume).[24]

The allusion in this quotation to the unchanging customs (and costumes) of the exotic world and the modern fashions splendidly displayed in the Palais (even adopting a French title to denote the latest mode in styles) suggests that, as much as racial or ethnic clues, clothing marked the players and audiences that mingled on the Pike. In the Palais du Costume, Western clothing, displayed from the Greeks onward, was arranged to denote a history, whereas native dress was described as either functional or ceremonial but generally unchanging. Dress probably helped to establish generally, as it did for Mrs. Wilkins, the distinctions that made entry into a strange and jumbled exotic world both adventuresome and safe for the American audiences. The "midway types" of ethnic peoples on display on the Pike and available in photographic reproductions fulfilled the expectation that they were others to be understood as representatives of the exciting, strange, and alien. Their clothing was defined as an unmistakable symbol of identity, worn in an unchanging tradition. This contrasted with the lavish gowns in the Palais, which celebrated the modern notion of clothing as style, or apparel that was changing and progressing. Here, dress was a disposable and individuated consumer item, not simply a necessity.[25] Just as portrait-type photography depicted the individual as a generic example, so clothing could define such an individual as representing a race or culture. As Thorstein Veblen had put it only a few years earlier in his *Theory of the Leisure Class,* modern dress was "an expression of the pecuniary culture." To the sociologist, clothing was designed to "make plain to all observers that the wearer is not engaged in any kind of productive labour." Unlike the stable costumes of the Japanese, he wrote sardonically, American styles were ever-changing demonstrations of waste.[26]

If Veblen exaggerated for polemical purposes, he was uncanny in offering a way to understand the encounters that occurred on the Pike between fashionably dressed Americans and the traditionally clad peoples employed by the concessions. But how, in this case, does this reflect Veblen's work-leisure distinction? Photographic evidence shows that most tourists on the Pike were dressed in the prevailing leisure-wear of the day: long skirts, shirtwaists, and broad-brimmed, plumed hats for women, and dark pants, white shirts, vests, and bowler or derby hats and caps for men.[27] To be sure, there were class distinctions in this American dress, visible most dramatically when officials of the Fair or visiting dignitaries appeared in black formal wear and top hats, with elaborate gowns for women. For those wearing native dress, their "work" was to appear "authentic," wearing clothing that identified them as a type or the employee of some concession. This work versus leisure distinction served further to divide spectacle and spectator. And even if native wear could itself be divided symbolically into complex styles designating class, age, and social grouping, it was probably difficult for the audience to recognize such separate categories.[28] As Georg Simmel wrote in his seminal essay entitled "Fashion," published in 1904 in the United States, "The savage is afraid of strange appearances; the difficulties and dangers that beset his career cause him to scent danger in anything new which he does not understand and which he cannot assign to a familiar category. Civilization, however, transforms this affectation into its very opposite."[29] Simmel further suggested that fashion served the important function of linking the individual to the group: "it furnishes a general condition, which resolves the conduct of every individual into a mere example," he wrote, but at the same time provided for "the desire for change." In contrast, there was no creative tension between continuity and change in uncivilized culture. Even in the dance, he wrote, "It has been noted quite generally that the dances of primitive races exhibit a remarkable uniformity in arrangement and rhythm." Clothing, in other words, belonged to a large category of cultural signifiers that differentiated the civilized from the savage.[30]

Fashion was thus a key ingredient in the self-confident visit of Fair patrons to the exotic quarters of the Filipinos and other exhibited peoples. If, as Kurt Back has written, style is "a combination of personal expression and social norms, influenced by dominant values," clothing had a special place in social communication. Whether consciously or unconsciously, Americans possessed the ability to read Western fashions as a code, designating varieties of cultural distinction such as age, gender, work and play, conformity and rebellion. The distinctions between the dressed and the undressed exaggerated and made obvious the overtones of power that fashion obliquely expressed.[31]

Another variant of dress clearly visible on the Pike and elsewhere in St. Louis was the uniform. Because of the large number and frequent rotation of military encampments at the Fair, uniforms were commonly seen. While

FIGURE 18. "Crowd in the East End of the Pike." This picture of the Pike suggests the size of the crowds that marched up and down through a cacophony of concessions. Even in this amusement section, the crowd of visitors was formally dressed. (Louisiana Purchase Exposition Company; Library of Congress Prints and Photographs.)

their uniforms might ordinarily be recognized as symbols of order, this was not always the case. Sometimes members of the militia congregated in "large bodies, and preceded by field music, and with much noise, to parade the 'Pike.'"[32] These parodies of ethnic and sideshow parades distressed concession owners, who feared the disruptions would drive away customers, and so they persuaded the administration to limit visits by men in military dress to more orderly planned and controlled tours. In addition, the Jefferson Guard, wearing its distinctive blue uniforms (and only lightly armed) attended to maintaining order and crowd control. If the Guard uniform symbolized established order, then the Philippine Scouts and the Philippines Constabulary Brass Band paraded in military uniforms that symbolized the aspiration of achieving order and civilization under the tutelage of U.S. occupation. This dress-up of dark bodies in American-style outfits created a similar impression to what struck visitors when they heard the Filipino band playing American patriotic songs and state anthems. They often expressed surprise and delight about the contrast and juxtaposition. Thus, the Kentucky Commission reported that on Louisville Day the first feature was the Igorot singing class: "This quartet of brown savages sang 'My Old Kentucky Home' in a way that surprised the entire audience." And as one newspaper noted about the parade marking the dedication of the Philippine Reservation, it included a striking contrast because of its "savage and military composition."[33]

Erotic dress along the Pike was also clearly an issue of attraction and concern. Knowing full well the sensational reputation of the amusement section at the Chicago World's Fair in 1893, the Board of Lady Managers issued a preemptive warning about potential scandal on the Pike. After all, self-appointed American morals censor Anthony Comstock had visited Chicago in 1893 to denounce the erotic *danse du ventre* (belly dancing) on the Midway and tried unsuccessfully to shut it down. At the initial meeting of this group of elite St. Louis women in October 1902, the first resolution passed by the board advised: "That it is the earnest desire of the Board of Lady Managers of the Louisiana Purchase Exposition, that there be no indecent dances or improper exhibits in the Midway during the Exposition; and that the Louisiana Purchase Exposition Company be urged to use the utmost care in awarding the concessions for shows, in order that there may be no objectionable features."[34] Their initial target was the Middle Eastern dancers that scandalized Chicago.

The most serious breach of propriety, though, had concerned the semi-nudity of the Igorots. Well aware that this attracted visitors to the Philippine Villages—and no one could miss the frequent newspaper articles that included lurid commentary on the Filipino body and its attraction for female visitors—the Board of Lady Managers was behind the move to force the Igorots into Western-style trousers. "War Department pants" were passed out, until President Roosevelt rescinded the order. The discarded trousers ended up on some of the other denizens of the compound and as decorations on "the roofs of the Igorrote huts."[35] Discussion of Filipino clothing bobbed up again over the course of the Exposition, from continued expressions of indignation over nakedness to worries that skimpy dress was inappropriate for chilly weather in the spring and fall.

The argument that finally prevailed defined nudity as culturally "authentic" among the Filipinos and, therefore, fully acceptable. University of Chicago anthropologist Frederick Starr actually asserted, "Dressed, they would lose their scientific value and might die." Even Anthony Comstock, who rushed to St. Louis to see for himself, pronounced the Philippine exhibit a triumph. As the *Ogden Standard* put it in mid-July, "Comstock spoke with admiration and enthusiasm of the beautiful copper colored physique of the Igorots, and saw nothing to which he could take exception in their semi-nakedness."[36] This coincided with the somewhat idiosyncratic anthropological viewpoint

FIGURE 19. The entrance to Creation. This vast and popular concession on the Pike shows another version of nudity, which excited no controversy when confined to classical motifs in statuary. (Official Photographic Co., 1904; Missouri Historical Society, Photographs and Prints Collections.)

of Albert Jenks. Writing for the *American Anthropologist* in December 1904 after the Fair, he noted that the Igorot clothing, such as it was, was entirely utilitarian in origin, lacking any of the usual reasons for costume, such as sexual attraction and eroticism.[37]

If, somehow, *authentic* could be construed as the opposite of *erotic*, then the seminaked bodies of the Filipinos could be (officially) defined as sexually benign. And it is true that the pitched conflict between Marines and the Scouts occurred while the latter were fully dressed in Western apparel. In other words, when Filipinos donned American-style clothing, they became a more plausible threat to transgress social boundaries. If we press this proposition further, we can see the operation of shifting and unequal definitions of erotic zones. Though Americans and Europeans sometimes experienced clothing as a form of erotic definition, calling attention in this period to ankles, wrists, necks, and other bare parts of the body, as well as the exaggerated buttocks and breasts of women slung forward into a fashionable S-curve, the experts at the Fair deemed the absence of clothing to be nonsexual.[38]

In addition, there is some evidence that the Filipinos themselves recognized the Western standards that defined clothing as erotic. In the record of an interview with Chief Antonio of the Igorots, the *Washington Post* reported that in response to a question about his nakedness, he retorted that to adopt Western-style clothing would be to don a "heritage of shame and disease that has sapped the vitality of society." The article reported that Antonio may not have understood all the implications of his argument, but that, in the long run, there was no answer to his wonder at Americans—and in particular the American lady in attendance at the interview—as to why such elaborate and uncomfortable clothing was deemed proper.[39] And yet we know that much of this argumentation was ultimately beside the point. Just as patrons sometimes neglected the finer distinctions of racial and cultural hierarchies and categories laid out before them, they (if not the anthropologists) clearly recognized the erotic possibilities in the seminudity of the native peoples displays despite the gloss of "authenticity" and flocked to see for themselves.

While the American press reported the pants incident with considerable amusement, there was, as with the entire imperialist venture in the Philippines, some recognition of the absurdity and cultural devastation of the endeavor. The poem "The Igorrote's Burden," published in the *Washington Post* in October 1904, portrayed the trousering in a mocking reprise of Kipling's famous imperialist poem:

> Take up the white man's burden
>   Ye Igorrote breed:
> Go, bind your sons in exile
>   In linen and in Tweed:
> Go, wait in heavy harness
>   Of shoes and hats and coats,
> Poor, new-caught, sullen people,
>   With Ascots 'round your throats.[40]

As the interview with "Chief Antonio" suggests, Filipinos themselves returned the Western gaze with a difference. A recent work by Jose D. Fermin on the Filipino experience at the 1904 Fair suggests the complex reaction in the island nation to the highly selective and in many respects degrading exhibition of native villages. Responding to the unrepresentative nature of the villages (chosen to emphasize primitivism) and the circumstances that emphasized barbaric qualities of Igorot culture, the Philippine legislature passed a law in 1908 forbidding the exhibition of tribal people unless fully clothed. While this law was not honored at subsequent exhibits, including the Panama-Pacific International Exposition in 1915 in San Francisco, it does suggest the angry reaction of educated Filipinos to displays of native peoples. There is also the suggestion in the astute introduction to the book that the experience at St. Louis and elsewhere helped bolster Philippine nationalism and a sense of identity. Toward the end, Fermin reports a visit by seventeen Igorot Americans to a sixth grade class in Clayton, Missouri (next to St. Louis), in May 2000. Held at the Wydown Middle School, which was sited on the spot once occupied by the Philippine Reservation, the school had, until the mid-1970s, nicknamed its sports teams and yearbooks The Igorots. One element of the event was an apology extended by the teacher and students for the derogatory mascot name, but in part it was a memory occasion that affirmed the presence of millions of Filipinos in the United States and their successful incorporation into American society.[41]

Another document that more imaginatively portrays the story of Filipinos at the Fair is the faux docudrama by Marlon Fuentes, "Bontoc Eulogy," produced in 1995. Based upon actual newsreel footage from the Fair and knowledge of Filipino Igorots, Fuentes creates the fictional story of Markod, a young man persuaded in 1904 to leave his wife for an adventure in the United States, and the filmmaker's own search to find traces of this, his invented grandfather. This retelling of the story of the Fair from the imagined perspective of an Igorot ancestor, replete with fictitious letters sent home to his wife, is the tragic narrative of someone wrenched out of an idyllic village culture and thrust into the human zoo of the Philippine Reservation.[42] It illustrates, in personal and poetic terms, the damaging illusions of McGee, Taft, and Roosevelt in their desire to illustrate the benefits of imperialism. Of greatest interest, however, is the creation of a new voice of memory, one that contradicts, or at least chastens, the dominant celebratory mode of Fair recollections. Whether it will modify the collective memory of the Louisiana Purchase Exposition is an important question relating to the virtual disappearance of the Philippine Reservation and its imperial implications from the bittersweet nostalgia that still dominates recollections of the Fair.[43]

For all the attention to the Pike and the Philippine Reservation, where the highest attendance seemed to occur, we need to remember that large numbers of patrons ventured elsewhere: to the great industrial and liberal arts and transport buildings, to the pavilions and palaces of other nations, and to the exhibits of the states of the Union. They wandered along the paved open spaces admiring the sculpture, gardens, and lagoons and watched the artificial Cascades by day and stayed late to see their illumination by night. If collective memory is accurate, they consumed copious amounts of ice cream in bent-waffle cones, purportedly invented on the fairgrounds that summer, and drank iced tea, another supposed innovation. They dined in restaurants and rode the intramural railroad. They visited countless small exhibits on the Pike and elsewhere. They listened to lectures, went to sporting events, and heard music. They attended special celebrations and commemorations, of which there were several every day marking the opening of exhibits or noting the visits of important dignitaries. They went to national conventions held on the fairgrounds and gathered together in large family reunions. In other words, they approached the Fair with what historians have called the perspective of "everyday life." We can surmise that their experience was an extension of what Dorothee Wierling calls the "quotidian life": in this case, the move from the known to the unknown.[44]

The Fair administration kept no formal attendance records for the various large exhibition halls, so we cannot accurately know which proved to be most popular in terms of numbers. Memoirs and newspaper accounts are, at best, impressionistic. But there are some fascinating hints that the Agriculture Building may have attracted the most visitors in the category of large exhibit halls. The final report of this exhibit included a spot-check estimate of the numbers in its building compared to visitors at other display palaces. The chief of the Department of Agriculture had "a very careful count made in all the exhibit palaces upon the grounds and in the Government Building." This was only an estimate of numbers present in the building at one time, but he found that for September and October, there were more visitors in Agriculture than in any other display space. In August, Agriculture was second only to Government. In fact, he reported, by these estimates, Agriculture more than doubled most other exhibits in terms of attendance. In the last three months of the Fair, "not less than thirty thousand visitors [were] in the Agriculture Building each day."[45] If that is so, then for each day during this period, approximately half of the visitors to the Fair entered the exhibit. This report may be self-serving, but it is supported by other evidence.

The *Waterloo Courier*, in August 1904, declared that late summer and fall (now that crops were in) would be the time for farmers to visit the Fair. They represented the "army of men who made its creation possible," the newspaper extolled. And agriculture was the most prominent department of the Exposition, it continued. Not only could rural Americans learn about new methods of crop rotation and view advanced implements and machinery, they could see themselves represented and experience a "new pride" of vocation and the "importance and dignity of their calling." Like so many other patrons at the Fair, farmers found a place and means to affirm their special identity in an important exhibit or concession.[46]

Oral histories suggest another reason why Agriculture might have been attractive to young patrons. As Mrs. Emile Waniger reminisced, the extravagant displays of fruit from California and Florida "intrigued me very much." Given the recent development of these subtropical agricultural areas and the abundance of fruits and vegetables they promised, this was obviously a wonder to a young child. As Willard Shelp remembered, there were endless treats and snacks available to sample: "cakes and cookies and different products that the different people in the food industry made that exhibited there and you could get a good meal and I did many many Saturdays."[47] For adult women, the array of food products and preserving suggestions were legion, and, since preparation was a key part of their domestic lives, these displays were probably as intriguing as the mechanical farm devices and livestock were for their husbands. We do not know the precise ratio of visitors from farms to more urban patrons, but this was a period of rapid movement from farms to small towns and small towns to larger cities. Many of these immigrants carried with them memories and close ties to agricultural life through extended families still living on farms. These agricultural displays may have been just as important and impressive to visitors as the steam turbine engines in the Machinery Building and, perhaps, visibly closer to the lives they lived.

What I will call exhibits and displays of proximity, which included the symbolism of home and daily life with the attraction of familiar identities, may have been as important as the exotic and unfamiliar sought out on the Pike and the Philippine Reservation. This footnote to the familiar may have amplified the dialectic between the known and unknown that I have argued was a constant experience of fairgoers. By seeing themselves celebrated on the fairgrounds as farmers, Germans, Irish, proud residents of states, leaders of the local elite, or members of clubs and organizations, visitors were perhaps more psychologically prepared to encounter the unknown. In fact, the connection between these two experiences seems to be a primary motive that guided their wanderings around the grounds between homelike and foreign places. As oral histories and memoirs reveal, fairgoers were constantly making comparisons between who they were and what they had already seen or known and what was on display.

If visits to the Irish or German concessions on the Pike or German and Irish displays in other parts of the Exposition reflected a hyphenated identity, we can also see this possibility reflected in the function of state pavilions that reminded visitors of their local affiliations. Architectural styles of these buildings attest to their significance as symbolic home sites. Plantation buildings for southern states, elegant colonial structures for New England states, and other architectural and historical footnotes served the same imaginative function as the Tyrolean Alps or Blarney Castle. In this case they conjured up the ideal of a state as a place of belonging. Anchored in history and geography, these state buildings were designed to attract two sorts of visitors: those having ties to the state by residence or travel, and those interested in immigration or investment.[48]

In fact, the state exhibits provided a remarkably broad range of services. As the *Report of the Maryland Commission* noted in 1906, the state recognized the importance of providing a "building a Marylander could call his own." The state even hired an official hostess, who lived in the building during the summer, making it "homelike" for Marylanders visiting the Fair. The structure also contained exhibit halls where Maryland products were displayed. In addition, the state mounted an exhibit in the Agriculture Building with its "Maryland Space," featuring census reports, real estate agent circulars, and other enticements to potential immigrants. This multiple presence was characteristic of other states that combined services for their home residents and potential immigrants and advertisements in other display buildings for products and investment.[49]

Bigger states like New York appropriated large sums for their buildings and exhibits. Practically a fair within the Fair, New York's exhibit attempted to illustrate the general theme "of processes, as well as of products." State literature boasted of its exhibits in education and social economy, such as prisons, lunacy care, and fingerprinting and other criminal identification tools. Besides welcoming New Yorkers, the state building hosted frequent receptions and banquets for visiting dignitaries and conventioneers. There were also frequent entertainments, such as minstrel shows.[50]

While the Exposition administration let no opportunity pass to celebrate the opening and dedication of an exhibit or concession or a visit by a foreign ambassador or national politician, the official openings of state exhibits were always celebrated with extravagant speeches and entertainments, visits by governors and state militias, and often the Philippine brass band playing the state anthem. Pennsylvania Day, for example, featured speeches by David Francis and the governor of Pennsylvania and a parade led by the Jefferson Guards, the Philippine brass band, the Philippine Scouts, U.S. Marines, a Mexican band, the Philippines Constabulary, the Philadelphia City Troop, and state dignitaries riding in open carriages.[51]

These were occasions to celebrate American patriotism but also to affirm the history of the state as it belonged to the ongoing project of creating the nation. As ex-President Grover Cleveland declared in his 1903 dedication speech for the Exposition, citizens should be "aroused to higher and more responsive patriotism by the reflection that it is a solemn thing to belong to a people favored of God." Thus, local history was integrated into a larger national patriotism. As the *New York Times* put it, "The Fair appeals to a particular and local patriotism in a region in which local patriotism is particularly rife." If this described the spirit of Missouri, it was true for other areas as well. States identified their historical place in the American march of progress. Thus, for example, Pennsylvania celebrated its symbolic identity as a founder of the nation when it shipped the Liberty Bell to St. Louis for exhibition. On the Fourth of July, the Pennsylvania Building recorded over twenty-five thousand visitors, and over a million for the duration of the Fair, many no doubt drawn by the display of this precious national icon in its state setting.[52]

Can we determine what visitors saw inside the major display palaces? How did visitors encounter the vast exhibits of the great exhibition halls? The overabundance of choice may well have determined what patrons did. For example, when they entered, patrons were confronted with a great array of objects and displays side by side inside cavernous spaces. The Fair management adopted a nomenclature designed to facilitate viewing and learning. Arranged in a kind of generic order, this created two possibilities. One could either follow the proliferation of objects arranged by type (looking at all railroad locomotives together, for example), or one could use the guide hanging at the entrance to locate a type of display or something specific suggested by an acquaintance, a guidebook or newspaper article, or one's individual interests. Given the limits of time, and probably only fitful interest in most objects, it is highly unlikely that visitors did anything more than encounter a few of these exhibits with much care. As diary and oral histories confirm, visits were disorganized and often did not seem to follow any prescribed order. In this sense the encyclopedic arrangement of the Fair had the opposite of the intended effect, making it possible and almost necessary to be highly selective.[53]

A description of what visitors actually saw as they traipsed around this immense exhibition is crucial, because modern aesthetics and display formats are quite distinct from what appeared at the turn of the century. A few contemporary observers were aware that the art of exhibition was changing, stimulated in large measure by new institutions such as department stores. As Marshall Everett wrote in 1904 in his *Book of the Fair*, "In the past few years this dressing of show windows has become a profession." Large emporia now hired staff to create "original and attractive installations."[54] Some of these display advances were apparent in individual exhibits, but the general environment was decidedly outdated and would have looked crowded to the modern eye.

Visitors encountered a variety of display genres: demonstrations of process and manufacture; scale models; consumer products; livestock; transportation equipment; art works; demonstrations of sociology, education, and anthropology; and human exhibits. In general, displays of goods were busy with

FIGURE 20. "Palace of Machinery Interior." This interior shot suggests how the vast floor of displays and the completely utilitarian interior design contrast with the neo-classical exteriors. (Official Photographic Co., Louisiana Purchase Exposition; Library of Congress Prints and Photographs.)

detail and dense explanatory text. Vast pyramids of canned goods and objects like oranges advertised, in extravagant fashion, the opulence of agricultural production, while giant animals contrived out of perishable fruit suggested the whimsy of abundance. Looking down the long rows of kiosks visible in photographs and stereographs, these displays appear overloaded, dwarfed by the immense beams and support columns, while the huge open upper spaces seem wasted.

Because items were organized by type of product or process rather than within any larger environment stressing their use, or the integration of an item within the home, for example, it was probably necessary for the visitor to imagine this setting. Where integration did occur, it was generally within an exhibit of some social process like an ideal classroom or in demonstrations of the application of science to everyday life, such as the cooking with electricity display in the New York State pavilion. As Mark Bennett wrote in his 1905 history of the Fair, the classification system allowed the viewer to investigate "any branch of humanity's progress or present condition" and to "consider the

FIGURE 21. Half of this stereograph of "Missouri Tomatoes" shows the seemingly endless range of vegetables on display. Viewed in three dimensions, the demonstration of abundance, so characteristic of the agricultural displays, was even more emphatic. (Library of Congress Prints and Photographs.)

whole series together, without having to mentally assemble them."[55] But, in fact, exactly this sort of mental integration was necessary. The typical arrangement can be seen in the organization of Group 53: Equipment and Processes used in Sewing and Making Wearing Apparel: Class 326—Common Implements used in needlework; Class 327—Machines for Cutting clothes, skirts, and leather; Class 328: Machines for Sewing; and so on. Wandering amidst all these choices must have been bewildering. Perhaps the administration recognized this problem by awarding prizes for the best objects and processes in each category. This, at least, would provide some indication of worth and identify a selection of displays to see. The administration appointed two hundred or so juries (about one-quarter were women), who bestowed thousands of separate medals given for a variety of accomplishments, although these didn't come until late in the summer.

Descriptions of this cataloging system stressed that it was "the most comprehensive and representative contemporary record of the progress and condition of the human race that may be devised. It is a succinct and authentic record of the exhibits of the Exposition, which are the evidences of the capacity and performance in all lines, intellectual, social and industrial, of all the peoples of the world."[56] If this universal scheme made it easy to locate a specific display, it also encouraged competition among manufacturers and nations. This was Edward Schneiderhahn's experience when he spent the afternoon in the Transportation Building. Comparing railroad locomotives built in several countries, he noted: "the American locomotive machinery far superior."[57]

Thus, the visitor was confronted with a vast collection of products and processes, with some guidance offered by the prize system but little else other than the attractiveness of the kiosk, the interest of the patron, some sort of prior knowledge or advice, or a guidebook entry to steer him or her. The result that we have seen from examining oral histories, diaries, and memoir literature was a disorganized sampling of displays, random visits that followed inclination, whim, or perhaps a group decision. At the same time, there is no indication that most patrons understood the larger point of the elaborate classification system or recognized why the Fair administration attempted to create a "universal encyclopedia." Like so much else at the Fair, visitors were obliged to transform what they saw and experienced into some appropriate individual meaning.

One lesser known function of the Fair was to serve as a convention center. Organizations made considerable use of the Fair, and a huge number of visits were experienced as a group event. For hundreds of local and national groups, the Louisiana Purchase Exposition became their site for a national convention or annual meeting. These organizations used the facilities of the fairgrounds to attract members from all over the United States. Military encampments, parades, and drills drew thousands of soldiers and militia to the grounds. The "Daily Official Program" listed their maneuvers as well as the many special conventions, dedications, and commemorative events that signaled the presence of special groups in St. Louis. For June 2, for example, the program mentioned the convention of the National Federation of Musical Clubs and a convention of American Press Humorists. The *World's Fair Bulletin* cited at least seventy fraternal association "days" when a group, meeting on the fairgrounds, received special public recognition. Guidebooks also remarked on these commemorations. As the *World's Fair Authentic Guide* put it, the October 6 German Day was expected by leading Germans throughout the nation to be "the most notable gathering ever held in this country."[58] There were also special event days dedicated to university alumni (of Michigan and Princeton, for example) and national fraternities, Christian Endeavor Day, and so forth. As many as 360 national organizations and groups held conventions on the fairgrounds, ranging from the Anti-Horse Thief Association, to the American Federation of Labor, the German Army and Navy Association, the Mexican War Veterans, the Philippine Veterans, the National Association of Manufacturers of the United States, the United Irish League, and the United Daughters of the Confederacy as well, of course, as the National Convention of the Democratic Party in July.[59] (Republicans declined to convene there.) In fact, so many groups chose St. Louis for their national meetings that it became practical to build a Temple of Fraternity financed through contributions by national fraternal organizations and women's groups. President Francis, in dedicating this building, enthusiastically addressed the importance of fraternal societies in the United States and their presence at the Fair. They were the "conservative element of our citizenship," opposing both anarchy and monopoly. He continued, "I have remarked that upon more than one occasion if I were asked to point upon these grounds to the building which of all others comes nearest typifying the spirit of this Exposition and the participation of all people in it, I would point to the Temple of Fraternity."[60] Even if Francis was inclined to speak in superlatives about every exhibit at the Fair, this remains a significant statement. In the end, the Exposition estimated that over 130,000 delegates attended the major congresses and conventions held at the Temple or in the much more commodious Music Hall and Festival Hall.[61]

There are several plausible explanations for this extraordinary convergence of conventions and national meetings at the Fair over the summer of 1904 in St. Louis, and for their role in affirming the "spirit" of the Exposition. Certainly the reduced fares offered by railroad companies were an enticement, as were the plentiful hotel rooms, restaurants, and meeting spaces on the grounds. And the Fair itself was a major draw in raising attendance and bringing members together.

Beyond this, however, the Fair, we must recall, was a place of ritualized nationalism, a space where groups and individuals could affirm the manner of their membership in the nation and recall their role in creating its history. From this perspective, the Louisiana Purchase Exposition was the modern American version of the medieval pilgrimage, in this case designed to assert a shared community in the midst of rapid change, just as that ancient ritual journey defined a dispersed Christian community. Fairs throughout this period opened up liminal spaces in which identity could be reimagined.[62] Ethnic groups like the Germans and Irish, religious organizations, states of the Union, the city of St. Louis itself, could all affirm, through their representation on the fairgrounds, a significant role in building the nation and a place in its future. The world's fair was not unique in this respect, for groups had long used local holidays and commemorations to assert their membership in the larger community. As Amatai Etzioni and Jared Bloom have noted, Americans frequently celebrated nationalism through its local manifestations.[63] The Louisiana Purchase Exposition was a highly sought-after and significant environment for this possibility, filled with symbols of national pride and history as well as footnotes to local patriotism.

The formal family reunions held on the grounds throughout the summer offered the same possibility to inscribe one's organization and one's identity into the national celebration of the Louisiana Purchase. These reunions were also enacted on a smaller scale by the informal visits of relatives to cousins, aunts and uncles, parents and children, or other family already residing in the city. Interest in family history and reunions had become widespread in American society during the late nineteenth century. In part, they gained a special significance and momentum after a time when the nation and many families were torn apart. In 1863 President Lincoln revived Thanksgiving as an annual holiday that made the commemoration of family togetherness a national ritual. The president's prayers for a restoration of the nation became an annual celebration that reassembled dispersed families. Formal family reunions began to become conspicuous in the 1880s and were well-known by the turn of the century, when their occurrence was frequently mentioned in newspapers.[64] Genealogical societies sprang up along with special interest heritage organizations like the Daughters of the American Revolution. By 1900 hundreds of genealogical and historical societies had been founded both in Europe and the United States.[65]

As Etzioni and Bloom write, Victorian family pilgrimages "to the secular shrine of the grandparents' house in the country" were a way to replenish the spirit before "returning to a more hectic life in a town or city."[66] First popular among Americans of Protestant heritage, these rituals soon spread to other groups. Periodic trips back to an ancestral home in the country were celebrated, especially in the southern part of the United States, although there were also frequent northern versions. These reunions reassembled kin and their spouses and offspring to participate in ceremonies of eating and family storytelling. As one historian has written of these events, three sentiments of Protestant tradition found their way to the heart of reunions. She writes that these include "themes of individuals leaving home for lands far away," of defending homes in the wilderness and doing battle with enemies of the faith, and "themes of a covenant community being established where the wickedness of the cities and towns can be shut out." They are, she concludes, rituals that honor both leaving home and the virtues of staying behind.[67]

Such meetings constructed a past that perhaps had never existed within a landscape of nostalgia that contrasted with the too real restlessness of modernizing society. As Susan J. Matt has written, "The ideal home was the old house, handed down from generation to generation, set in a rural landscape, undisturbed by modern life." This celebration was a counterpoint to the growing late nineteenth-century interest in vacations on Vermont farms and other rural idylls.[68]

Family reunions characteristically returned to the nostalgic place of family origins: a farmstead or a small town. To relocate such a meeting within a city and especially in an imaginary and transitory urban environment like a world's fair indicates a dramatic shift in meaning or, perhaps better, an assertion of meaning and belonging within a new context. In part, reunions were made more practical by the pull of the Exposition and the cheap rail fares and hotel rates of St. Louis. But holding a reunion on the fairgrounds also endowed this momentous family occasion with fresh significance as it transposed the sanctified homestead and individual identity to an imaginary representation of the nation and its future.

Reunions could be elaborate events with official sanction, and there were many recognized by the Fair administration in the official events program. For example, the minutes of the Committee on Ceremonies disclose plans for the Paxton Family Reunion Day in late June of 1904. The Paxton "society" arranged for hotel reservations at the local Christian Endeavor Hotel for a three-day get-together, but requested a formal meeting space on the grounds on the twenty-ninth. "The family is to be organized at this time," continued the minutes, "and they have a record of 500 who will be here, and a genealogy of 50,000 names, all indexed, to which they will send out invitations as soon as their date is determined."[69] A family reunion, even if not on such a grand scale, gave extraordinary meaning to the contemporary popular song "Meet Me in St. Louis, Louis." Here was a place to complete the circles of belonging: to a family, ethnic group, organization, section, state, nation, race, and, finally, to civilization itself.

In such ways, individuals reconstructed the Louisiana Purchase Exposition to fit their own purposes, interests, and ideas. Their effort to understand the vast display with its complex schematic and endless distractions, all overlaid with the heavy ideology of race and imperialism, meant translating these ideas into their own vocabularies. We know from reader-response theory as well as the studies of museum behavior that individuals find their own, often idiosyncratic meanings, even in the best planned of events. The Fair management seemed well aware of this possibility. As a writer for the *Architectural Review* noted in 1904, entrance locations were placed so as to oblige visitors to enter and make a turn that would present the fairgrounds from its most impressive perspective. In this way the planners tried to correct a fault perceived at the prior exposition in Buffalo, where patrons avoided the main entrance and "drifted in at all corners."[70] Yet drift they did, anyway, pulled by a variety of contending impulses and most of all by the sheer variety and scale of the exhibition. Despite the overtly didactic intentions of many displays, the lure of novelty, the exotic, and the entertaining was just as strong, suggesting that the Exposition was used as much as an amusement park as the encyclopedia of civilizations, more an emporium than a university. This appears increasingly to be the case with late-Victorian world's fairs. The Concession Report for the St. Louis Fair concluded that (with prices more or less constant) patrons spent considerably more money on amusements at the Louisiana Exposition, somewhat more than $1.00 per person per visit, compared to only $.75 at Chicago just ten years before.[71] In fact, the general history of world's fairs moved increasingly to exploit commercial and entertainment possibilities.

We also need to speculate whether this willful behavior might also have its analogue among those groups that did not attend the Fair personally but gained information from a wide variety of published sources. Could the same be said for those visitors who prepared themselves in advance by reading newspaper stories, guidebooks, and journal articles? If these sources helped pique interest in the oddities of the Exposition, as I have discussed in the case of the Philippine Reservation, they also called attention to the architectural splendors of the Exposition, its fascinating concessions, and its displays of modern machinery and consumer items. The publicity department of the Exposition worked conscientiously to distribute long, detailed accounts that were sometimes republished in their entirety as news articles. Did these accounts influence what patrons saw at the Fair? Surely they did, although how much we cannot know. But can we legitimately assume that a newspaper story or extended journal article was read or understood in its entirety, that readers received the message intact?

Reader-response theory suggests that all communication between authors and audience is dependent upon a complex interaction between text and individual. Interpretation and meaning seem to depend upon a community of interpretation—that is, the conventions which the reader applies to what he or she is reading or, in the case of photography, viewing. While this conclusion may suggest anarchic versions of significance, it matches our understanding of the way patrons at the Fair interacted with the displays set before them. Here too, individuals imposed their own agendas and itineraries. They frequently made sense of the didactic lessons in anthropology and empire within the context of their own racial ideas and language.[72] In the end, they endowed the Fair with a significance that suggests that these late-Victorian fairs were public spaces where individuals could act out the ways in which they belonged to the nation. Perhaps this belonging was tenuous; perhaps American identity required frequent affirmation. Perhaps it was more fragile, more transitional than we have supposed. A trip to the Fair might be a journey to find oneself in a world of multiple possibilities and disconcerting change.

CHAPTER SEVEN

MAKING IDENTITIES

Sometimes the meaning of an event is best understood by those looking in from the margins, those excluded for one reason or another, who could not attend but recognize the importance of being present. This was certainly the case for the National Association of Colored Women's Clubs, which planned to hold their biannual convention on the fairgrounds in July 1904. The sad story of this futile endeavor and the group's intense desire to affirm membership for middle-class African Americans in the larger American community by appearing on the fairgrounds underscores the importance of the Louisiana Purchase Exposition as a site of modern pilgrimage. It was a family reunion of the nation they wanted to attend. To be a recognized part of that family, their strategy was to establish an identity more middle class and more mainstream than the mainstream itself. And still they felt excluded.

Their self-imposed exile was the inevitable reaction to the reputation for discrimination of the Exposition, for its unfair treatment of African Americans in restaurants and other concessions. Black newspapers across the country repeated the story of the Eighth Illinois Regiment, which had planned an encampment on the grounds but had to cancel when the administration demanded that it set up on separate grounds and supply its own tents and commissary. This was the result of insistence by southern-state military groups, who refused to share space and a mess hall with black soldiers.[1]

Initially, David Francis had hoped to diffuse problems of segregation by avoiding them. As a memo on a proposed "Bureau for Negro Race" noted,

"Every effort has been made to avoid raising the issue [of race]." Negotiations were held to establish a Negro Day (or Emancipation Day) on August 1 on the fairgrounds. A local committee of black St. Louis citizens was set up to do the planning and applied successfully to the Committee on Ceremonies of the Louisiana Purchase Exposition Company for permission. Plans included a series of addresses, a chorus of a thousand voices, and a civic and military parade (the latter led by the Eighth Illinois Regiment). We can imagine from the scale of this planning what was at stake. In a letter inviting Booker T. Washington to attend, the chair of the local committee, W. M. Farmer, spelled out the strong reasons to be present: "No better place could be chosen than this city," he wrote, "where will be in progress the greatest Exposition the world has ever witnessed. The nations of the earth will be assembled here, the attention of the world will be centered here. Aside from the wholesome influence of well prepared addresses and strong resolutions, the coming together of so many thoughtful men and women of the race can not fail to make a favorable impression on the assembled multitude."[2]

But hopes for Negro Day were dashed when unfavorable publicity about segregation on the grounds reached the national press. As chair of the local committee, Farmer wrote to Booker T. Washington again, this time rescinding his invitation: "We suppose it is well known to you that the sentiment throughout the country is to the effect that 'the Negro is not wanted at the World's Fair.'" To be honest, Farmer continued, that was not entirely true, although there had been a few incidents, but the unfavorable reputation would undoubtedly turn away potential delegates and visitors. So the celebration was cancelled.[3] While Francis had personally opposed segregation and discrimination on the fairgrounds, he had also acceded in several instances to practices that repelled potential black patrons, and this was the result.

The National Association of Colored Women's Clubs (NACWC) actually met in St. Louis in July but, in an abrupt reversal, changed the venue for its only session scheduled for the fairgrounds. Instead, the group met that day at the St. Louis Colored Orphans Home in the city. This withdrawal was approved after a resolution early in the convention tendered by Mrs. [Margaret Murray] Booker T. Washington, denouncing the Fair for bias in hiring and for tolerating a general pattern of discrimination. Even though Francis had appointed a "colored woman" to represent, meet, and greet black patrons at the Fair and offered free passes to editors and reporters from black newspapers, this unfavorable reputation was too strong to ignore.[4]

The cancellation is even more poignant because the early history of the NACWC is intimately connected to the history of American world's fairs and their racial policies and practices. The first black women's clubs were organized in the early 1890s to fight against waves of increasing segregation and lynching. Ida B. Wells, the energetic Memphis journalist and newspaper editor, was instrumental in creating the first national club organization. But the national organization was also spurred into being as a specific response to the segregation and exclusion of African Americans from the Columbian Exposition in 1893 in Chicago.[5]

The NACWC held its first convention in Boston in 1895 to coincide with a large conference of the Christian Endeavor Society, to which many of its members also belonged. Future meetings were generally set to coincide with large public gatherings: Nashville in 1897 during the Tennessee Centennial Celebration and Buffalo in 1901 for that city's world's fair. In fact, the group postponed its biennial meeting for a year to gather during the Louisiana Purchase Exposition in St. Louis in 1904, partly, as J. S. Yates in the office of the president put it in 1902, "when we know we can get exposition rates to St. Louis." If successful, the group understood this would be an immensely important time to "show to the world as we could not in any other way, the 'Progress of Negro Womanhood.'"[6]

To be present at such events and witness for the centrality of the black women's club agenda in American culture was a fervent desire of the organization. Past success at the Pan American Exposition in Buffalo, where the mayor of the city addressed the group, made the Louisiana Purchase Exposition potentially even more important since Missouri was a border state. All the more significant, then, were the slights sustained in St. Louis. As a result, the group did not again meet in a city simultaneously hosting a world's fair or major exposition until 1933, when, once again, the organization skipped a year in its biennial schedule to come together in conjunction with the Century of Progress Exposition in Chicago.[7]

When the three hundred or so women did finally assemble in St. Louis, the program of their conference was an extraordinary enactment of patriotism, elite culture, middle-class manners, and a fastidious observation of order and decorum. The group passed strong resolutions denouncing both lynching and segregation, but it also recorded its approval of the Women's Christian Temperance Union agenda and passed several declarations elaborating on the belief "that the mother is the rock upon which the home is built." Addresses affirmed the need to instill middle-class values among black families. The minutes to the July 13 evening session describe a speech by Miss Anna H. Jones, president of the Missouri State Federation of Women's Clubs, as a sermon on uplift and propriety. "This address," the minutes recorded, "was replete with lofty thought, important facts and statistics bearing upon the condition and progress of the colored woman in America, discussion of the odds against which she has had to work in her upward struggle from bondage to civilization, refinement and culture; and timely suggestions relative to her future progress."[8]

This stress on manners and uplift—from reiteration of the slogan "Lifting as We Climb" to the performance of classical music, poetry readings by a "world-renowned" elocutionist, to the singing of traditional Protestant hymns—all recorded an elaborate symbolism that expressed a potent yearning to belong and be recognized for the remarkable progress made in reshaping former slaves into upright citizens since Emancipation. Mrs. Josephine B. Bruce, chair of the Executive Committee, made a plea for "a better home-life, better schools, better churches, sober living, and a purer moral atmosphere."[9] It also seemed necessary to reject elements of black culture which appeared vulgar. Thus, Mrs. Ida J. Jackson of the Department of Music reported on the aim to develop a taste for classical music "and in words of no uncertain sound, [she] denounced 'coon songs and rag-time music.'" The absence of any performance of spirituals, what W. E. B. Du Bois had identified as the "Sorrow Songs," underscored this striving to hone an identity that would look identical to the mainstream. Indeed, the group might have fit well onto the Board of Lady Managers, clucking at sexual improprieties on the Pike. Thus, they knocked at the doors of cultural equality and were rebuffed. They imagined a community that could not imagine them.

This distressing failure to achieve insider status at the Fair underscores the centrality of the identity performance that can be found in almost every other element of the Exhibition. Their rejection only highlights the imperative of belonging. This was something experienced as crucial for those visitors who held family reunions on the grounds. For those who registered at their state pavilions, enjoyed romanticized versions of the old country and American history, and who asserted a common identity by observing the cultural and ethnic and racial distinctions visible everywhere at the Fair, this was a momentous occasion. As the members of the National Association of Colored Women's Clubs realized, the Fair provided the possibility to affirm the many layers of American nationalism. It offered a way for individuals to understand on their own terms and in their own vocabularies the modern revolution in consumer goods and technology, a world shrunk by imperialism into near neighborhoods of different cultures and ethnicities, something echoed in the growing diversity of the United States itself.

In this respect, the Louisiana Purchase Exposition performed familiar and not uncommon cultural work. It gave Americans another place, in this case an imaginary environment, in which to overcome (or evade) some of the great divisions in society posed at the end of the century. These included the slow-to-knit wounds of the Civil War, the entrance of the United States into the dangerous and controversial scramble for empire, the continuing divisions of race, a burgeoning multiethnicity in culture, rapid abandonment of the farm and urbanization, the end of the Indian Wars, and the amazing, westward trek of white population. It occurred at a moment when, as Matthew Jacobson argues, the United States was reacting to and absorbing millions of new immigrants even as it embarked on a defining imperial adventure in the Philippines.[10] The Exposition offered up an environment where visitors could make sense of the confusing and divided modern world. Not unlike the elaborate urban pageants and other celebrations and commemorations of the day, it attempted the same stock-taking affirmations, only on a grander scale, in an era that could not seem to invent enough occasions to enact this experience.

In the beginning of this book I asked what we could learn from history and memory and what might be changed by paying closer attention to the experience of visitors to the Fair. Historians have lucidly depicted the intentions of the Fair administration and many of the exhibitors. The emergent collective memory of the Fair places the event at the center of St. Louis history and the identity of the city, cleansing recollection with the soft soap of local pride and nostalgia. I have also shown that visitors transformed the well-crafted intentions and lessons of administrators and exhibitors into contexts of their own understanding, putting the new and exotic into familiar terms, sometimes even playing with new identities—and suspending the rules that prevailed outside the fairgrounds. I have discussed how important the pleasure principle was in guiding the visitor's engagement with the Fair, threatening sometimes to turn it into a vast country fair or sideshow. If historians have helped us understand the ideology that guided the Fair's construction, memory has affirmed that the Fair as an event lost these ideological markers and increasingly became the symbol of a disappeared past, replete with the unfulfilled promises of a city whose best moment may well have been in 1904. Experience, however, asks us to revisit both of these conclusions and demands to know how visitors participated in these gigantic pilgrimages to sites where they insisted on leaving a trace of their own citizenship. They were, after all, the ones who gave meaning to this event. As they still do.

At the intellectual congresses, despite Hugo Munsterberg's determination and despite the effort to look forward, the meetings summed up the present by looking backward at an intellectual world that would shortly face disorienting challenges. Although many of the industrial displays featured examples of advanced technology and invention, their meaning and impact had not yet become apparent at the Fair, and only became such in retrospect. As far as can be determined, visitors rarely reacted to them except as consumer items, or understood them in the context of their own lives and occupations. Many of the impressive technological achievements were employed primarily to enhance entertainment. The most mentioned "inventions" of the Fair were hot dogs, ice cream cones, and iced tea. This is entirely consistent with the testimony of visitors to the Fair that emphasized entertainment.

Looking at experience insofar as we can determine it, what, finally, can we conclude about how to write and—perhaps better—rewrite the history of the St. Louis Louisiana Purchase Centennial Exposition and similar events? What can we say about the project of history writing and memory preservation in general as it concerns such public events? Historians, as well as those interested in preserving the memories of early St. Louis, inevitably reproduce their own concerns into interpretations. And they should. Necessarily this means that today's issues of race and anticolonialism must constitute a major part of our assessment. Focusing on the intentions of the Fair planners rather than the reactions of visitors, historians have described carefully designed and pervasive racial and imperial ideologies. This is certainly not wrong, but evidence suggests that visitors made their own sense out of these ideas and translated them into their own vocabularies. There is little to suggest after the Fair that the categories of savage, barbaric, civilized, and enlightened entered into everyday racial and ethnic discriminations or replaced the fault line of white and black in American culture.[11] Does this mean that what patrons saw or did not see is irrelevant to the construction of meaning? Should history evaluate the perceptions of contemporaries? Obviously, I believe it important to reconsider their role. As I have tried to show, taking experience seriously weaves new patterns of interpretation around such events, dramatically changing our conclusions. In the end, we might assert that meaning represents our understanding of the dialectic between intention and experience. But I should stress that it is *our* understanding and *our* reconstruction of intention and experience. Entering into the minds and experiences of others is a Sisyphean task of rolling a boulder up the mountain of objectivity.

The reader will recognize by now that I believe the St. Louis Exposition to constitute many fairs, representing multiple experiences. As the Division of Press and Publicity understood, the Fair, in terms of its audience, was primarily a local event with national and international aspirations. Those who designed the Fair themselves had several purposes: some narrowly expressed in ambitions for the city and state economy, some national and even imperial. They celebrated and advocated (at the same time) a modern St. Louis. They proposed an organization of knowledge that would fit the world's peoples into a hierarchy of order that would make sense of the immense variety of peoples, cultures, and economies that imperialism had made proximate. At the same time, they sought to integrate knowledge and display into several familiar contexts, in the clichés of contemporary concession culture. In some respects St. Louis was a local agricultural fair, but also a major contributor to the burgeoning culture of world's fairs. It shared many aesthetic elements with amusement parks and department stores. It built a modern city as perfect and orderly as the most optimistic Utopian of the day could hope for. All of these models were apparent at St. Louis, as they were and would be at other fairs.

Those in attendance clearly contributed to the meaning as much by what they appeared not to notice as by what they constructed on their own. Their experience was manifestly rooted in place and occupation, ethnicity, social class, and gender. A significant determinant of this experience centered on the family, either as the group that attended together or as part of the occasion to visit with relations in the city. St. Louis itself figured prominently in the Fair in expressions of ethnic origins, city and state loyalties, and, finally, nationalism. These converging circles of belonging helped construct the context in which visitors understood what they were seeing and what they later remembered.

The enormously successful German Day celebration presents a good example of this sort of case. The fourth largest day for attendance (behind St. Louis Day, Labor Day, and Closing Day), this celebration of German American identity was marked by what Audrey Olson has described as a parade of forty thousand men "each carrying an American flag." Contemporary newspaper accounts tell of the huge crowd sporting white ribbons saying "Do Your Duty to the Old Country and the New." If this is perhaps an exaggerated description, it nonetheless suggests the importance of enacting this hyphenated identity on the fairgrounds in the presence of invited German dignitaries, American officials, and German American performing groups of singers and athletes.[12]

Finally, it should be emphasized that the amusement section played a leading role in the way meaning was established and promoted at the Fair even if the administration did not set out with this purpose. As John Fiske has written, "Popular culture is made by the people at the interface between the products of the culture industries and everyday life."[13] If we transpose this notion to comprehend the reaction of visitors to the various cultural narratives presented to them, we can better understand how they might have reacted to the immense variety of possibilities. This is particularly important because the amusement section offered popular versions of the serious displays mounted elsewhere. Thus, the Pike presented an entertainment version of anthropology, just as the Boer War reenactment presented imperialism as theater and spectacle. Commerce and entertainment were rampant at the Indian Reservation, as figures such as Geronimo charged a fee for his photograph and other souvenirs. Ethnic origins in the guise of the Tyrolean Alps presented a fantasy memory of the fatherland for thousands of German residents of St. Louis and an imaginary Europe for other Americans.* Modern science in the guise of the Baby Incubator was a popular concession. Along the Pike in the concession depicting The Old Plantation, American slavery and race relations were reenacted for the sake of nostalgia and an entrance fee.[14] The exhibits of "savage peoples" and exotic cultures blurred the distinctions between anthropology and entertainment.[15] Religion, in the well-attended Creation concession, was reshaped into an amusement. Indeed, almost all of the most serious portions of the Fair had their analogue in some form of commerce, commodity, or entertainment. Even the amazing electrical technology involved in nighttime lighting was memorable primarily as a showy splendor. These were the things experienced and remembered. Patrons responded most enthusiastically when contemporary social issues and inventions, and even modernism itself, could be transformed into some form of entertainment. The energy and fun of amusements created an immediately usable present that would later become a usable past. Entertainments were extensions, translations, perhaps even parodies, of the most serious elements of the Exposition. Chicago's Paul Blackmer understood the fundamental truth about the fairs when he suggested that the audience would act as it pleased, and if it were pleased, its acts would make the Fair a success.

This is where attention to experience can be generally useful for historians. Perhaps the most important conclusion we can draw is that the visitor to the Fair encountered the modern world not just intellectually, as an encyclopedia of knowledge, but as a spectacle expressed within the logic and conventions of amusement. This made it possible to possess the Fair and its exhibits in congenial terms. As Mrs. Wilkins demonstrated, dancing across the dangerous boundaries of race and ethnic differences in an atmosphere of suspended rules could be a joyous and carefree affair. We might even suggest that the idiosyncratic ways visitors related to the Fair, what they chose to see and where they carefully spent their few dollars, can tell us something about how the nation as a whole responded to the Fair. Perhaps the rest of America looked at photo books and stereographs and read the literary and newspaper accounts from similar perspectives and personal angles, imposing their own interpretations and meanings, picking and choosing what to understand and remember, rather than absorbing everything that was laid out for them.[16]

---

*The Bavarian Alps at this time was a part of the highly contested Tyrol region of Germany, Austria, and Italy.

Michel de Certeau, in his influential book *The Practice of Everyday Life*, suggests that consumption of culture rarely reproduces and internalizes the intentions of those who create it. Indeed, he argues that the process is inherently one of resistance. Thus, the ingenious ways that the weaker members of society make use of rules and order create a "political dimension to everyday practices."[17] This notion of resistance by ordinary people is rightfully understood as a way of understanding the elaborate dynamic of culture. In de Certeau's argument, there is also the suggestion that this sort of empowerment has a progressive element to it. Perhaps this is true in many cases, but that is not my argument. The multiple meanings that Americans derived from the Fair and their experience with its displays were not necessarily enlightened, particularly in terms of racial issues. What is important is the degree to which visitors and observers encountered the Fair as an extension of their own lives and practices.

What then is left to say about the elaborate itineraries developed by the planners of the Fair? These itineraries and intentions were, of course, present in the displays and the publicity that poured from the administrative offices of the Louisiana Purchase Exposition Company. Articles on the evolution of races filled American journals and newspapers; and photographs, books, postcards, and other media reproduced the vision of primitivism and alien civilizations. But it cannot be assumed that their arguments about race or imperialism completely carried the day. Theirs was a hotly contested anthropology that would eventually become a discredited anthropology under the influence of such scientists as Franz Boas and Margaret Mead. Nor was imperialism in the innocent and celebratory guise displayed at the Fair really victorious as an ideology. Indeed, the Philippine acquisition never ceased to be controversial, and, under President Woodrow Wilson, just a few years later, American foreign policy resumed its traditional position of opposing blatant forms of European imperialism and colonialism. And it is just as likely that most observers fit their vicarious encounters with other cultures into the framework of prevailing racial ideas rather than adopting the complex hierarchies of McGee. As an exasperated Abraham Lawshe reported on the Philippine Reservation, "no amount of emphasis on commercial exhibits, constabulary drills and scout parades has distracted attendance from the 'dog-eaters' and 'head-hunters.'"[18]

Was this the world's greatest fair of all—as the collective memory of St. Louis has recalled it? Considered in the sober light of revised attendance figures, it fell short of Chicago in the 1890s and certainly well below later fairs of the 1930s. Like all fairs in this period and later, it was instrumental in demonstrating some of the possibilities of modern technology and busy with defining the consumer citizenship to which Americans increasingly paid allegiance. But so did other sorts of expositions and displays and, indeed, so did cities themselves with their multiplicity of new institutions like museums, amusement parks, skyscrapers, businesses, factories, and modern entertainments.

The Fair was a distinct genre of public amusement and learning. It offered possibilities for enacting the practical rituals of belonging to modern society within the context of an all-embracing environment of modern entertainments, ideas, products, and processes. Much of what patrons saw and experienced was freely available elsewhere: in the booming cities of New York, Chicago, and St. Louis. It could be witnessed in amusement parks, civic parades, department stores, urban parks, city streets, immigrant ghettos, factories and farms, shows, entertainments, and other forms of tourism. But nowhere was the totality of vision so accessible. And nowhere else was it so easy to affirm and celebrate one's own place within this emergent world than in the multiple venues celebrating identity offered by the Fair.

Fairs presented what seemed both unproblematic and condensed, purging the turmoil, violence, and messiness of modern life that visibly accompanied the transformation to modernity. Fairs were, to use Michel de Certeau's striking metaphor, attempts to render the labyrinth of the city transparent and legible.[19] St. Louis summed up much that was known about the world intellectually and culturally, but it intentionally ignored almost as much. When we finish our calculations, we can surely conclude that the Louisiana Purchase Exposition marked a great moment in the history of the city—its high point, even. But in the larger scheme of American history, it was only one fascinating effort among many to understand and control the rapid and disorienting changes that were happening everywhere in society at the same time. By analyzing it as a case study, where these forces were intensified, we can learn a great deal about how Americans at the time understood their changing world—how they invested their identities in glimpses of a future that they would help shape and make commonplace.

NOTES

INTRODUCTION

1. For a discussion of some of the anxieties and clashes between historians and keepers of memory, see Ian Tyrrell, *Historians in Public: The Practice of American History, 1890–1970* (Chicago: University of Chicago Press, 2005).
2. Harry Harootunian, *History's Disquiet; Modernity, Cultural Practice, and the Question of Everyday Life* (New York: Columbia University Press, 2000).
3. David Lowenthal, *The Past Is a Foreign Country* (Cambridge: Cambridge University Press, 1985), 192. Lowenthal quotes Eugene Minkowsky in this passage.
4. Clarence L. Cullen, "Riot of Megaphones," *Washington Post*, October 9, 1904.

CHAPTER ONE

1. The *Encyclopedia Britannica* is an exception, listing accurate paid admissions, although with no attempt to disaggregate them. Frederick Pittera, "Exhibitions and Fairs," *Encyclopedia Britannica* (Chicago: William Benton, 1972), 8:961.
2. *New York Times*, August 20, 1939, 34. The article on the New York fair, reporting on George Gallup's questionnaire to patrons, indicated that each person went an average of 2.3 times. Martha Clevenger also suggests that there were multiple visits to the St. Louis fair but makes no effort to estimate the actual attendance. Martha Clevenger, ed., *"Indescribably Grand": Letters and Diaries from the 1904 World's Fair* (St. Louis: Missouri Historical Society Press, 1996), 17.
3. "Recapitulation of Monthly Totals of Admissions from April 10 to December 1, 1904, and Free Admissions on Sundays," in "Final Report of the Department of Admissions," Louisiana Purchase Exposition Company Papers, box 13, folder 1, n.p., Missouri Historical Society, St. Louis, Missouri. Hereafter cited as LPE Co. MSS.
4. "Notes of a Conversation," December 10, 1902, box 13, series v, subseries i, folder 2, 203, LPE Co. MSS. The Street in Cairo exhibit was among the most popular exhibits in Chicago but less so in St. Louis.
5. Ibid., 5–6. Blackmer's name in the report is spelled two different ways. I have chosen "Blackmer" as the most plausible.
6. I arrive at a figure of 5,500,000 by adding 500,000 (Chicagoans) to Blackmer's estimate of 5 million outsiders. The St. Louis Fair sold season passes. These accounted for 352,391 adult admissions, with the average number of entries being over 90 per pass. Each booklet had 184 tickets. Department of Admissions, "Final Report of the Department of Admissions," February 2, 1905, box 13, series v, subseries 11, file 1, 28, LPE Co. MSS. Children's coupon book admissions amounted to only about 4 percent of total passbook entries.
7. Department of Admissions, "Final Report of the Department of Admissions," February 1, 1905, 27–28, 31.
8. John Findlay, *Magic Lands: Western Cityscapes and American Culture after 1940* (Berkeley: University of California Press, 1992), 80. Disney also found that most of the patrons were from

California, with lesser numbers from west of the Mississippi. They also belonged, generally to the middle- and upper-middle-income groups, 89.

9. Walter Benjamin, "Paris, Capital of the Nineteenth Century," *Reflections: Essays, Aphorisms, Autobiographical Writings* (New York: Harcourt Brace Jovanovich, 1978), 151.
10. Neil Harris, "Expository Expositions: Preparing for the Theme Parks," in *Designing Disney's Theme Parks,* ed. Karal Ann Marling (Paris: Canadian Centre for Architecture, 1997), 19.
11. These are designated by the Bureau International des Expositions according to a variety of criteria.
12. *Glimpses of the Lewis and Clark Exposition, Scenic Views* (Chicago: Laird & Lee, 1905).
13. Jennifer Crets, "What the Carnival Is at Rome, the Fair Is at St. Louis: The Nascent Years of the St. Louis Agricultural and Mechanical Fair," *Gateway Heritage* 22, no. 4 (2002): 24–33. Francis quoted in *Massillon Independent,* July 29, 1887. Francis had close ties with Cleveland, later serving as his secretary of the interior.
14. John Kasson, *Amusing the Million: Coney Island at the Turn of the Century* (New York: Hill & Wang, 1978).
15. "They Buckle Antlers On," *Waterloo Daily Courier,* June 16, 1899. (This is the story of an Elks Clubs convention in St. Louis.)
16. Mary Ryan, *Civic Wars: Democracy and Public Life in the American City during the Nineteenth Century* (Berkeley: University of California Press, 1997), 224, 302.
17. David Glassberg, *American Historical Pageantry: The Uses of Tradition in the Early Twentieth Century* (Chapel Hill: University of North Carolina Press, 1990), 67, 159–62.
18. Thomas M. Spencer, *The St. Louis Veiled Prophet Celebration: Power on Parade, 1877–1995* (Columbia: University of Missouri Press, 2000), 67, 164. The author notes that the annual parades represented peoples of the world in the order of least to most "civilized."
19. Walter B. Stevens, *St. Louis, One Hundred Years in a Week: Celebration of the Centennial of Incorporation, October 3, 1909* (St. Louis: St. Louis Centennial Association, 1909). Stevens lists some typical themes for the parades: "The first year the Creation was pictured in moving illuminated tableaux. Then came the progress of Civilization, the Four Seasons, A Day Dream of Woodland Life, Around the World, Fairyland, The Return of Shakespeare, Arabian Nights, American History, History of the Bible," 79. The 1904 parade was the first to use electrically illuminated floats.
20. Ralph Davol, *A Handbook of American Pageantry* (Taunton, MA: Davol Publishing, 1914), 17. See also Naima Prevots, *American Pageantry: A Movement for Art and Democracy* (Ann Arbor: University of Michigan Research Press, 1990), 1.
21. Davol, *Handbook of American Pageantry,* 105.
22. Prevots, *American Pageantry,* 9–14. See also David Glassberg, *Sense of History: The Place of the Past in American Life* (Amherst: University of Massachusetts Press, 2001), 68–71. Black groups in St. Louis, excluded from this pageant, organized the Veiled Prophet Mystic Order in 1905 as a counterbalance to the all-white performance. The St. Louis Veiled Prophet Organization was finally integrated in 1979. John A. Wright, *Discovering African American St. Louis: A Guide to Historic Sites,* 2nd ed. (St. Louis: Missouri Historical Society Press, 2002), 28.
23. James Neal Primm, *Lion of the Valley: St. Louis, Missouri, 1764–1980* (St. Louis: Missouri Historical Society Press, 1998), 362.
24. David Francis, speech, "At the Mayor's Dinner, Saint Louis," p. 1, box 16d, series ix, folder 3, p. 1. In his dedication-day speech, President Theodore Roosevelt celebrated this bully walk across the continent: "Never before had the world seen the kind of national expansion which gave our people all that part of the American continent lying west of the original states," he exclaimed. Speech quoted in David Francis, *The Universal Exposition of 1904,* 2 vols. (St. Louis: Louisiana Purchase Exposition Co., 1913), 1:143.

25. Chicago had a considerably higher percentage of immigrants than St. Louis at this time, about 30 percent to about 20 percent.
26. There was an existing Fairground Park that had long been the host of fairs, races, and celebrations, including the Veiled Prophet commemorations, but Forest Park was deemed more suitable.
27. Lara L. Johnson, "Cheap Thrills and the New Urban Order: Forest Park Highlands," *Gateway Heritage* 1 (1995): 70–81. The park closed permanently in 1963 after a fire.
28. *World's Fair Bulletin* 5 (May 1904): 34–35.
29. Harvey Levenstein, *Seductive Journey: American Tourists in France from Jefferson to the Jazz Age* (Chicago: University of Chicago Press, 1988), xi, 157–75.
30. Frederick J. V. Skiff, director of exhibits, preface to *Official Catalogue of Exhibitors: Universal Exposition, St. Louis, U.S.A., 1904, Department D, Manufactures*, by Louisiana Purchase Exposition, Division of Exhibits, rev. ed. (St. Louis: Official Catalogue Co., 1904). This preface appeared in all the official divisional catalogs.
31. M. H. Hulbert, introduction to *Official Catalogue of Exhibitors: Department D, Manufactures*; and Frederick J. V. Skiff, "Official Report, Division of Exhibits, Louisiana Purchase Exposition Company, 1904," box 5, xiii, folder 1, p. 1, LPE Co. MSS.
32. *Official Catalogue of Exhibitors: Department D, Manufactures*.
33. David R. Francis, "Attractive Features of the St. Louis Exposition," *Century Magazine* 68 (June 1904): 264, 268.
34. *Official Catalogue of Exhibitors: Department G, Transportation* (St. Louis: Official Catalogue Co., 1904).
35. Howard J. Rogers, "The History of the Congress." See also [no author], "Purpose and Pace of the Congress"; and Hugo Munsterberg, "The Scientific Plan of the Congress," all quoted in *International Congress of Arts and Sciences*, ed. Howard J. Rogers (Boston: Houghton, Mifflin, 1906), vol. 1: 1, 13, 27, 50, 92. Munsterberg was explicitly echoing Auguste Comte's classification of knowledge, 100ff.
36. A. W. Coats, "American Scholarship Comes of Age: The Louisiana Purchase Exposition, 1904," *Journal of the History of Ideas* 22 (July 1961): 404–17.
37. Turner's second effort fell far short of his brilliant articulation of the Frontier Thesis in 1893. Frederick Jackson Turner, "Problems in American History," and Duncan Macdonald, "The Problems of Mohammedanism," both in *International Congress of Arts and Sciences*, ed. Rogers, vol. 2: 527, 567. Munsterberg was distressed that so few volumes were printed and sold.
38. Philippine Exposition Board, William P. Wilson, chair, *Official Catalogue, Philippine Exhibits: Universal Exposition, U.S.A., 1904, Section D* (St. Louis: Official Catalogue Co., 1904), 7, passim.
39. Albert Ernest Jenks, quoted in *Official Catalogue, Philippine Exhibits*, 261.
40. *Official Catalogue of Exhibitors: Universal Exposition, St. Louis, U.S.A., 1904, Department of Anthropology* (St. Louis: Official Catalogue Co., 1904), 1–2. For a discussion of the dying-race narrative and its relationship to colonization, see Patrick Wolfe, "History and Imperialism: A Century of Theory, from Marx to Postcolonialism," *American Historical Review* 102 (April 1997): 388–420.
41. Diane Coombes, *Reinventing Africa: Museums, Material Culture and Popular Imagination in Late Victorian and Edwardian England* (New Haven, CT: Yale University Press, 1994), 65.
42. Vanessa R. Schwartz, *Spectacular Realities: Early Mass Culture in Fin-de-siecle Paris* (Berkeley: University of California Press, 1998), 1, 6–16, 135–37.
43. *World's Fair Bulletin* 4 (February 1904): 6, 3. June 4 was commemorated with a "Parade of Peoples and Beasts."
44. *World's Fair Bulletin* 3 (October 1901): 5.

45. W. J. McGee et al., "Report of the Department of Anthropology," typescript, box 30, series iii, subseries xi, p. 26, LPE Co. MSS.
46. John Troutman and Nancy J. Parezo, "The Overlord of the Savage World: Anthropology and the Press at the 1904 Louisiana Purchase Exposition," *Museum Anthropology* 22, no. 2, 17–21.
47. An example of continuing elaboration of racial differences can be seen in Madison Grant, *The Passing of the Great Race or the Racial Basis of European Civilization* (New York: Scribner, 1916). Grant's book sales topped a million and a half copies and had a long life, well into the 1930s. An example of anthropological refutations of contemporary racial theory can be found in the later works of Franz Boas and his students.
48. Harper Barnes, *Standing on a Volcano: The Life and Times of David Rowland Francis* (St. Louis: Missouri Historical Society, 2001), 135, 111.
49. "Very Queer Things Are to Be Seen on Every Side: Curious Odds and Ends," *Atlanta Constitution*, May 8, 1904.
50. *The Library and Its Work* (Washington: Government Printing Office, 1904), 11–13, 5.
51. *World's Fair Bulletin* 4 (January 1904): 40.
52. Both poems were published in *World's Fair Bulletin* 4 (May and June 1904): 55. The tone of this doggerel, and specifically the phrase "future Great," may be a reference to newspaperman Logan Reavis, who wrote a book on St. Louis in 1870 by that title. Reavis, who began the *American Tribune*, also suggested relocating the U.S. capital to the Mississippi Valley. Logan Urial Reavis, *Saint Louis: The Future Great City of the World* (St. Louis: St. Louis County Court, 1870).
53. "Transcript of a Conversation between Paul Blackmer and Norris Gregg," December 10, 1902, box 13, folder 2, LPE Co. MSS. See also "Minutes of a Meeting of the Concessions Committee," December 9, box 13, folder 3, LPE Co. MSS. This includes an extensive discussion about the content of exhibits and the revenue possibilities for a "Cairo to Stampol" concession owned by Mr. Pangalo.

CHAPTER TWO

1. Francis, "Attractive Features of the St. Louis Exposition," 264, 268.
2. Francis, *Universal Exposition of 1904*, 1:312.
3. Steffens's *McClure's* articles were republished in his 1904 *Shame of the Cities* (1904; New York: Hill & Wang, 1992).
4. Ibid., 522.
5. Barnes, *Standing on a Volcano*, 135.
6. Marshall Everett, *The Book of the Fair* (St. Louis: Henry Neil, 1904), 343.
7. Ibid., 325–26, 373.
8. Alfred Newell, *Philippine Exposition: World's Fair St. Louis, 1904* (pamphlet), 1.
9. Thomas R. MacMechen, "A Ten Million Dollar Pike and Its Attractions," *The Piker* (May 1904): 5.
10. Walter B. Stevens, "Report of the Acting Chief, Division of Press and Publicity," box 11, series iv, subseries ii, folder 1, pp. 4ff., LPE Co. MSS.
11. Division of Press and Publicity, "The Climax of Effort: Report," box 11, series iv, subseries ii, folder 1, 27, LPE Co. MSS.
12. Special issue, *Cosmopolitan* 38 (September 1904). The August issue also featured a number of articles on the Fair, stressing many of the same subjects: advanced technology, the anthropology exhibit, the Pike, and education. *Cosmopolitan* 37 (August 1904).
13. "The World's Fair at St. Louis," special issue, *World's Week* 8 (August 1904): 5057ff.
14. Albert Kelsey, "A Municipal Exhibit," *Architectural Review* 11 (July 1904): 188; William Lewellyn Saunders, "How to See the Fair," *Independent* 56 (May 19, 1904): 1130–34.

15. Mabel Loomis Todd, "The Louisiana Purchase Exposition," *Nation* 78 (June 30, 1904): 510–11.
16. "Closed," *Nation* 79 (December 8, 1904): 451; "The Cultural Features of the St. Louis Exposition," *Nation* 79 (December 22, 1904): 499.
17. Walter Williams, "Round the World at the World's Fair," *Outlook* 77 (August 6, 1904): 794–803.
18. Francis, "Attractive Features of the St. Louis Exposition," 265, 267–68.
19. The Philippine Scouts was the military force organized by the United States in 1901. The Philippines Constabulary was a national police force organized in the same year. Representatives of both groups were present at the Fair.
20. "Press Humorists Are Amused at World's Fair by Barbarous Dance of the Savage Igorrotes," *St. Louis Globe-Democrat*, June 1, 1904. The Missouri Historical Society Archives contain a number of volumes of scrapbooks with newspaper clippings which illustrate the points of this chapter.
21. Hermann Knauer, *St. Louis and Its World's Fair* (Berlin: G. Bernstein, 1904), 5–7, passim.
22. Eli J. Sherlock, *Sherlock's World's Fair Guide and Bureau of Information* (Kansas City: Sherlock, 1904), 4.
23. W. S. Wrenn, *The Rand-McNally Economizer: A Guide to the World's Fair, St. Louis, 1904* (Chicago: Rand, McNally, 1904), 7, 25ff.
24. Laird and Lee, *Laird and Lee's Standard Pocket Guide and Time-Saver* (Chicago: Laird & Lee, 1904), 85.
25. The exact wording of this slogan appears to be "processes as well as products," although it was often varied and paraphrased.
26. W. W. Ellis, *The Piker and World's Fair Guide* (St. Louis: National Publishing Co., 1904).
27. Official Guide Company, *World's Fair Authentic Guide: Complete Reference Book to St. Louis and the Louisiana Purchase Exposition* (St. Louis: Official Guide Co., 1904), 20, passim.
28. Ibid., 92.
29. See, for example, the photographic book *The Universal Exposition: Beautifully Illustrated* (St. Louis: Official Photographic Co., 1904).
30. Charles M. Stevens, *Uncle Jeremiah and His Neighbors at the St. Louis Exposition* (Chicago: Thompson & Thomas, 1904), 296–313, 32.
31. Ibid., 309. Stevens's next Uncle Jeremiah book, at the Panama Pacific Exposition of 1915, depends less on dialect and has less detailed description of the fair. Charles M. Stevens, *Uncle Jeremiah at the Panama-Pacific Exposition* (Chicago: Hamming-Whitman Co., 1915).
32. Jane Curry, *Marietta Holley* (New York: Twayne, 1966), 7–72.
33. Josiah Allen's Wife (Marietta Holley), *Samantha at the St. Louis Exposition* (New York: G. W. Dillingham Co., 1904), 290.
34. Ibid., 275–90.
35. Mark Bennett, *History of the Louisiana Purchase Exposition* (New York: Arno Press, 1976), xi–xiii, 465.
36. Coats, "American Scholarship Comes of Age"; Pittera, "Exhibitions and Fairs," 8:956–63.
37. There was some discussion earlier of holding the Olympics in Chicago, but St. Louis was chosen because of the Fair.
38. Primm, *Lion of the Valley*, 340–95. Stephen J. Raiche also stresses the city improvements stimulated by the Fair, "The World's Fair and the New St. Louis," *Missouri Historical Review* 67 (October 1971): 98–121. See also Edo McCullough's *World's Fairs Midways* (New York: Arno Press, 1976). As the title suggests, this work stresses the themes of continuity and innovation through each of several amusement sections of expositions. See also Dorothy Daniels Birk, *The World Came to St. Louis: A Visit to the 1904 World's Fair* (St. Louis: Bethany Press, 1979). This work is difficult to categorize and, like several similar works, rests at the borderline between history and a work of memory and celebration.

39. Burton Benedict et al., *The Anthropology of World's Fairs* (London: Scholar Press, 1983), 3–59.
40. Neil Harris, *Cultural Excursions: Marketing Appetites and Cultural Tastes in Modern America* (Chicago: University of Chicago Press, 1990).
41. "Report of the Director of Exploitation, Foreign Exploitation," box 11, series xiv, folder 2; and "Report of Domestic Exploitation," box 11, series iv, folder 1, LPE Co. MSS.
42. Robert Rydell, *All the World's a Fair: Visions of Empire at American International Exhibitions, 1876–1916* (Chicago: University of Chicago Press, 1984), 183, passim; and Robert Rydell, *World of Fairs: The Century-of-Progress Expositions* (Chicago: University of Chicago Press, 1993), 22, passim. While I do not agree entirely with Rydell's assessment of the Fair's meaning, I am deeply indebted, as are all historians of international expositions, to his groundbreaking work.
43. Beverly K. Grindstaff, "Creating Identity: Exhibiting the Philippines at the 1904 Louisiana Purchase Exposition," *National Identities* 1, no. 3 (1999): 245–63.
44. Jane Cutler, in her 1998 novel *The Song of the Molimo* (referring to a musical instrument from the Philippines), rewrote the history and fate of Benga in a novel that explored the Pygmy's reaction to American racism. (New York: Farrar, Straus & Giroux).
45. Phillip Verner Bradford and Harvey Blume, *Ota: The Pygmy in the Zoo* (New York: St. Martin's, 1992).
46. See Alan Trachtenberg, *Shades of Hiawatha: Staging Indians, Making Americans, 1880–1930* (New York: Hill & Wang, 2004), xxiii, 212.
47. Troutman and Parezo, "Overlord of the Savage World," 17–34. In an unpublished work, Martin Rudolph Brueggemann contests the central point of the historians who identify anthropology as an aggressive science in its infancy, seeking to convince the public of its new insights into human difference. Instead, he argues that such racial theories already prevailed, at least among ethnologists. Moreover, visitors were more likely to be attracted by the scandal of nudity and dog-eating at the Philippines compound than by an imperialist ideology or racial demonstration. The Fair thus appealed not to education in the sense intended by the Fair directors, but by entertainments on the Pike. This new world of consumption, unguarded by Victorian restraints, was far more important than the encyclopedic displays of industrial process and education. And if the Fair changed visitors at all: "They walked away more willing to participate as social actors in a society geared for consumption and recreational leisure." Martin Rudolph Brueggemann, "St. Louis' 1904 World's Fair: A 'Thick' Exposition" (BA thesis, Reed College, 1987), 149.
48. Paul A. Kramer, "Making Concessions: Race and Empire Revisited at the Philippine Exposition, St. Louis, 1901–1905," *Radical History Review*, Winter 1999, 107. Kramer extended this article in his book *The Blood of Government: Race, Empire, the United States and the Philippines* (Chapel Hill: University of North Carolina Press, 2006).
49. Ibid., 79–80, 94–102.
50. Nancy Parezo and Don D. Fowler, *Anthropology Goes to the Fair: The 1904 Louisiana Purchase Exposition* (Lincoln: University of Nebraska Press, 2007), 47–48, 73ff, 166–67, 324, 299. What the authors do not investigate is whether or not the message put forth was, in fact, received.
51. William Everdell, *The First Moderns: Profiles in the Origins of Twentieth Century Thought* (Chicago: University of Chicago Press, 1997), 206–26.
52. Eric Breitbart, *A World on Display, 1904: Photographs from the St. Louis World's Fair* (Albuquerque: University of New Mexico Press, 1997). Timothy J. Fox and Duane R. Sneddeker's book *From Palaces to the Pike: Visions of the 1904 World's Fair* (St. Louis: Missouri Historical Society Press, 1997) also presents an extensive photographic program to illustrate what is essentially a much less sharply critical reconstruction of the Exposition and a collection more focused on surveying the vast variety of displays, exhibits, and daily events. Robert Jackson's *Meet Me in St. Louis and a Trip to the World's Fair* (New York: HarperCollins, 2004) is a popular account of the Exposition and its marvels. It particularly emphasizes children. At the same time, this work

includes a long discussion of discrimination against African Americans and the involuntary appearance and display of native peoples, Filipinos, and other "primitive groups gathered for viewing and study," 99 ff.

53. Henry Adams, *The Education of Henry Adams* (1918; Cambridge, MA: Riverside, 1946), 466.
54. Warren I. Susman, *Culture as History: The Transformation of American Society in the Twentieth Century* (New York: Pantheon, 1973), 269.

CHAPTER THREE

1. Eric Breitbart, *A World on Display* (Corrales, NM: New Day Films, 1994), documentary.
2. Paul Ricoeur, *Memory, History, Forgetting* (Chicago: University of Chicago Press, 2004), 351.
3. Pierre Nora, "Between Memory and History: Les Lieux de Memoire," *Representations* 26 (Spring 1989): 9.
4. See David W. Blight's fascinating discussion of memories around the American Civil War for a discussion of the contest between history and memories. Blight, *Beyond the Battlefield: Race, Memory, and the American Civil War* (Amherst: University of Massachusetts Press, 2002).
5. Nora, "Between Memory and History," 7–24. See also Noa Gedi and Yigal Elam, "Collective Memory—What Is It?" *History and Memory* 8 (June 30, 1996).
6. Jacques Le Goff, *History and Memory* (New York: Cambridge University Press, 1992), 1, xi, 129.
7. John R. Gillis, "Memory and Identity: The History of a Relationship," in *Commemorations: The Politics of National Identity* (Princeton, NJ: Princeton University Press, 1994), 3–20.
8. Barbara A. Misztal, *Theories of Social Remembering* (Maidenhead, Berkshire, England: Open University Press, 2003).
9. John Bodnar, *Remaking America: Public Memory: Commemoration and Patriotism in the Twentieth Century* (Princeton, NJ: Princeton University Press), 14. A sea of ink has been spilled in the culture wars surrounding this issue. See, for example, E. D. Hirsch, Jr., *Cultural Literacy: What Every American Needs to Know* (Boston: Houghton Mifflin, 1987); Allan Bloom, *The Closing of the American Mind: How Higher Education Has Failed Democracy and Impoverished the Souls of Today's Students* (New York: Simon & Schuster, 1987); Lynne V. Cheney, *American Memory: A Report on the Humanities in the Nation's Public Schools* (Washington, DC: National Endowment for the Humanities, 1987); and Gary Nash, *History on Trial: Culture Wars and the Teaching of the Past* (New York: Knopf, 1997).
10. Michael Frisch, "American History and the Structures of Collective Memory: A Modest Experience in Empirical Iconography," *Journal of American History* 75 (March 1989): 1130–55.
11. On the difficulties of confirming the stories of Holocaust survivors, see the discussion of traumatic memory and its problematic reliability in Mark Roseman, *A Past in Hiding: Memory and Survival in Nazi German* (New York: Henry Holt, 2001).
12. Roy Rosenzweig and David Thelen, *The Presence of the Past: Popular Uses of History in American Life* (New York: Columbia University Press, 1998).
13. George Lipsitz, *Time Passages: Collective Memory and American Popular Culture* (Minneapolis: University of Minnesota Press, 1990), 5, 23. See also Dena Elizabeth Ebner and Arthur G. Neal, *Memory and Representation: Constructed Truths and Competing Realities* (Bowling Green, OH: Bowling Green State University Popular Press, 2001).
14. David Lowenthal, "The Timeless Past: Some Anglo-American Historical Preoccupations," *Journal of American History* 75 (March 1989): 1277.
15. Alison Landsberg, *Prosthetic Memory: The Transformation of American Remembrance in an Age of Mass Culture* (New York: Columbia University Press, 2004), 18, 47, passim.
16. On the instability and contentious meanings of memory, see especially, Jaclyn Jeffrey and

Glenace Edwall, *Memory and History: Essays on Recalling and Interpreting Experience* (Lanham, MD: University Press of America, 1994).
17. Endel Tulving and Fergus I. M. Craik, *The Oxford Handbook of Memory* (Oxford: Oxford University Press, 2000). See also Donald A. Ritchie, *Doing Oral History: A Practical Guide* (New York: Oxford University Press, 2003).
18. Alessandro Portelli, "What Makes Oral History Different," in *The Oral History Reader*, ed. Robert Perks and Alistair Thomson (London: Routledge, 1998), 78. See also Howard Schuman, Robert F. Belli, and Katherine Bischoping, "The Generational Basis of Historical Knowledge," in *Collective Memory of Political Events: Social Psychological Perspectives*, ed. James W. Pennebaker, Dario Paez, and Bernard Rime (Mahway, NJ: Lawrence Erlbaum, 1997), 73–76.
19. Ricoeur, *Memory, History, Forgetting*, 147.
20. See, for example, Theodore P. Wagner, "Year of the Fair," *St. Louis Post-Dispatch*, May 23, 1954.
21. Lillian Schumacher, "World's Fair Diary," Missouri Historical Society, St. Louis.
22. Beverly Bishop, "Analysis of Transcription," Louisiana Purchase Oral History Collection, box 1, file 2, Missouri Historical Society. Hereafter cited as Oral History Transcripts.
23. "Diary of Edward Schneiderhahn," in *Indescribably Grand*, ed. Clevenger, 41–50.
24. Ibid., 53–71. Clevenger does not comment on the differences between the diary and the memoir.
25. *Memories of the Louisiana Purchase Exposition, St. Louis, Mo., 1904: Containing over 125 Views of Buildings and Grounds in Halftone* (Grand Rapids, MI: James Bayne Co., 1904).
26. Francis, *Universal Exposition of 1904*, 1:312, xiv–xv.
27. Records of meetings: February 24, 1916; March 11, 1910; and April 29, 1910, folder 1, in Louisiana Purchase Exposition Historical Association Papers, Missouri Historical Society, St. Louis. Hereafter cited as LPE Historic MSS.
28. "Minutes of a Meeting for the Establishment of a Museum," box 24, series xii, folder 6, pp. 2–5, LPE Co. MSS.
29. W. J. McGee to A. C. Stewart, president of the Saint Louis Public Museum, January 1, 1906, Letters of the Board of Directors, Papers of the St. Louis Public Museum, Missouri Historical Society, St. Louis.
30. "Resolution of the Board of Directors, April 29, 1910," box 24, series xii, folder 3; and "Minutes of the Louisiana Purchase Historical Association, May 26, 1916," box 26, series xiii, folder 2, LPE Historic MSS.
31. David Francis, "Remarks," April 30, 1911, folder 1, LPE Historic MSS.
32. Michel-Rolph Trouillot, *Silencing the Past; Power and the Production of History* (Boston: Beacon, 1995), 53. The author also notes, quite rightly, that much of the history produced outside the academy is ignored by professional historians.
33. "Minutes of the Annual Meeting," April 30, 1911, folder 1, LPE Historic MSS. After the meeting, the group attended a movie presentation.
34. Sally Benson, *Meet Me in St. Louis* (New York: Random House, 1942), 269.
35. Ibid., 270, 212.
36. Perhaps there is no other explanation than the huge influence of the Catholic Church in Hollywood during this era for the strange appearance of two Sister of Charity nuns in the final frame of the movie. The scene is the only one of the Fair itself; two nuns move slowly across the screen before the Smith family appears. To employ them to symbolize the audience at the Fair is jarring and inappropriate.
37. Garland was a well-known performer on Armed Forces Radio at the time.
38. *Braukoff* is the Americanized spelling of *Braukhoff*, of apparent Prussian origin. There are other elements in the film that conjure wartime. For example, there is the enormous excitement of Rose

about a long-distance call from her romantic interest. And, no doubt, the Christmas song that stressed being "together again" contrasted vividly with the separated families of World War II.
39. There are some interesting visual parallels between the two films in the blue-and-white costumes worn by Judy Garland. Baum was a newspaper man in Chicago and a budding novelist when he wrote *The Wonderful Wizard of Oz*, first published in 1900.
40. Christopher Sergel, playwright and dramatizer, composed a play based on the Benson stories using the same name as the memoir and film. Sergel greatly altered the narrative, inventing two new characters associated with Mr. Smith's business, and focused more on the story of the job transfer to New York. He also restored the Waughop name and made Mrs. Waughop a busybody and nosey neighbor. The Fair still hovers as a grand coming event, but not, I think, as important as in the film. There is little, if anything, that echoes the war in this meager adaptation.
41. Larry Schulenburg, "The World Came Here 42 Years Ago; It May again in 1953," *St. Louis Globe-Democrat*, July 7, 1946, 1.
42. "Interview with Eric Link," *St. Louis Post-Dispatch*, April 29, 1979, 16.
43. "Women of Visitors Center in the Old Post Office Will Recreate Famous Exposition on April 28," *St. Louis Globe-Democrat*, April 15–16, 1967.
44. Department of Parks, "A Fair to Remember," press release for "A Fair to Remember, 1904–79," scrapbook 1, Missouri Historical Society, St. Louis. See also "A Fair to Remember," calendar of events, ibid.
45. Department of Parks, schedule of events for "A Fair to Remember, 1904–79," ibid.
46. See, for example, Sharra L. Vostral, "Imperialism on Display," *Gateway Heritage* 13, no. 4 (1993): 18–31; Johnson, "Cheap Thrills and the New Urban Order"; Michael Lerner, "Hoping for a Splendid Summer," *Gateway Heritage* 19, no. 3 (1998–99): 28–41; and Crets, "What the Carnival Is at Rome." The Historical Society also publishes *Voice*, an online version. The Missouri Historical Society also has on its Web site a virtual, interactive fair, based on the exhibit "Looking Back at Looking Forward."
47. Richard Friz, *Official Price Guide to World's Fair Memorabilia*, 1st ed. (New York: House of Collectibles, 1989), 25. For collectors of exclusively St. Louis memorabilia, see Robert L. Hendershott, *The 1904 St. Louis World's Fair: Mementos and Memorabilia* (Iola, WI: Kurt R. Krueger, 1994).
48. The group claimed to have about 100 charter members and 276 members in 1987.
49. Gregory Freeman, "St. Louis, Meet Me at the Fair," December 29, 1989, quoted in *World's Fair Bulletin* (1904 World's Fair Society) 5 (January 1990): 4.
50. "1904 Fair Receives Unfavorable Press in New Biography," *World's Fair Bulletin* (1904 World's Fair Society) 7 (October 1992): 3.
51. There is considerable dispute about the origins of this name, with many observers arguing that it is much older than 1904. See the series in the *St. Louis Post-Dispatch*, September 11–13, 16, 1986.
52. *World's Fair Bulletin* (1904 World's Fair Society) 19 (May 2004).
53. Randi Korn and Associates, "A Summative Evaluation of 'Meet Me at the Fair: Memory and History and the 1904 World's Fair,'" prepared for the Missouri Historical Society, May 1997, appendix, 1–4, copy lent by Katharine Corbett. See also Jeffrey and Edwall, *Memory and History*.
54. Katharine Corbett and Howard S. Miller, "Meet Me at the Fair: Memory, History, and the 1904 Louisiana Purchase Exposition," NEH Proposal, November 28, 1994, copy supplied by Corbett.
55. Ibid., 6
56. Ibid., 7–12.

57. Katharine Corbett, "How Exhibits Mean," manuscript copy; published in *Exhibitionist* 18 (Fall 1999).
58. Ibid., 11.
59. Katharine T. Corbett and Howard (Dick) Miller, "A Shared Inquiry into Shared Inquiry," *Public Historian* 28 (February 2006): 36.
60. For a thorough discussion of these issues, see John H. Falk and Lynn D. Dierking, *The Museum Experience* (Washington, DC: Whalesback Books, 1992).
61. Randi Korn and Associates, "Summative Evaluation of 'Meet Me at the Fair.'" This fascinating study reveals a very unrepresentative audience attending the exhibit. In terms of racial categories, about 87 percent were white, 3.5 percent black, 3.5 percent Native American, and 1.4 percent Asian American. This does not accurately reflect the demographic profile of the city of St. Louis and its surrounding area, although it may reflect the way in which memory "belongs" to one ethnic group or another, 15.
62. The city held a celebration of one hundred years of the Olympics in 2004.
63. Clevenger, *Indescribably Grand*, 21–23.
64. Interview with Ernest Link, box 1, file 3, Oral History Transcripts.
65. Interview with Louis Venverloh, box 1, file 34, Oral History Transcripts.
66. Interview with Willard Shelp, box 1, file 28, Oral History Transcripts.
67. Siegfried Kracauer, "Photograph," in *The Mass Ornament: Weimar Essays* (Cambridge, MA: Harvard University Press, 1995), 50.
68. There is a marked and fascinating equivalence between the singular, sacred objects of collective memory—such as the original Liberty Bell, for example, or the Fort McHenry flag from 1812—and souvenirs belonging to individuals, which may well have monetary value based upon their identification with events celebrated by collective memory.

CHAPTER FOUR

1. Division of Press and Publicity, "Climax of Effort: Report," folder 1, 18–19, LPE Co. MSS.
2. Sam Hyde, "Scrapbook of 1904 World's Fair, Recollections of the Fair," Sam Hyde Album, Louisiana Purchase Exposition World's Fair Albums, Missouri Historical Society, St. Louis.
3. Susan Sontag, *On Photography* (New York: Farrar, Straus & Giroux, 1973), 3–15.
4. Ibid., 65–67, 156, 165.
5. Roland Barthes, *Image-Music-Text* (Glasgow: Fontana/Collins, 1977), 31–44; Sontag, *On Photography*, 108.
6. John Tagg, *The Burden of Representation: Essays on Photographies and Histories* (Minneapolis: University of Minnesota Press, 1993), 4–5, 61–63; Kracauer, *Mass Ornament*, 47–63.
7. Peter Burke, *Eyewitnessing: The Uses of Images as Historical Evidence* (Ithaca, NY: Cornell University Press, 2001), 10ff. See also Ronald E. Doel and Pamela M. Henson, "Reading Photographs: Photographs as Evidence in Writing the History of Recent Science," in *The Historiography of Contemporary Science, Technology, and Medicine: Writing Recent Science*, ed. Ronald E. Doel and Thomas Soderqvist (London: Routledge, 2006): 201–36.
8. Lenz albums, Louisiana Purchase Exposition Collection, Missouri Historical Society, St. Louis.
9. White City Art Co., *Snap Shots of the Saint Louis Exposition, 1904* (Chicago: White City Art Co., 1904), 2.
10. C. S. Jackson, *Jackson's Famous Photographs of the Louisiana Purchase Exposition* (Chicago: Metropolitan Syndicate Press, 1904).
11. *Memories of the Louisiana Purchase Exposition.*
12. *The Universal Exposition: Beautifully Illustrated* (St. Louis: Official Photographic Co., 1904).

13. Walter B. Stevens and William H. Rau, *The Forest City: The Official Photographic Views of the Universal Exposition Held in Saint Louis, 1904* (St. Louis: N. D. Thompson, 1904).
14. William Rau, Director of Photography, *The Greatest of Expositions Completely Illustrated* (St. Louis: Official Photographic Co., 1904). Rau also published a number of stereo sets, including one depicting the Spanish-American War.
15. Alexander Alland, Sr., *Jessie Tarbox Beals: First Woman News Photographer* (New York: Camera/ Graphic Press, 1978), 45.
16. The Library of Congress, for example, has a collection of photographs by the Gerhard sisters. See also Anne Maxwell, *Colonial Photography and Exhibitions: Representations of the "Native" and the Making of European Identities* (London: Leicester Press, 1999), 85, 105, 123–27.
17. Frances Benjamin Johnston, "Negrito Archers, Philippine Reservation," Library of Congress, image no. LCj6962.
18. World's Fair Album, Louisiana Purchase Exposition Manuscripts, Missouri Historical Society, St. Louis. And St. Louis World's Fair Photograph Album (multivolume), Missouri Historical Society, St. Louis.
19. Richard Carline, *Pictures in the Post: The Story of the Picture Postcard and Its Place in the History of Popular Art* (Philadelphia: Deltiologists of America, 1972), 12, 53–62.
20. Robert Bogdan and Todd Weseloh, *Real Photo Postcard Guide: The People's Photography* (Syracuse, NY: Syracuse University Press, 2006), 4–30.
21. Rosamond B. Vaule, *As We Were: American Photographic Postcards, 1905–1930* (Boston: David R. Godine, 2004), 19–23. 52.
22. Robert Stumm, *A Postcard Journey Back to Old St. Louis and the 1904 World's Fair* (Springfield, IL: Octavo Press, 2000), 242.
23. Erin C. Blake, "Zograscopes: Virtual Reality and the Mapping of Polite Society in Eighteenth-Century England," in *New Media, 1740–1915*, ed. Lisa Gitelman and Geoffrey B. Pingree (Cambridge, MA: MIT Press, 2003), xix, 1–4. Oliver Wendell Holmes, *The Stereoscope and Stereoscopic Photographs* (New York: Underwood & Underwood, 1906), 9, 34. Note that Holmes's enthusiastic essay on stereographs was published by one of the leading manufacturers of stereo cards: Underwood & Underwood.
24. Harold F. Jenkins, *Two Points of View; The History of the Parlor Stereoscope* (Elmira, NY: World in Color Productions, 1957), 20–22. William Culp Darrah, *Stereo Views: A History of Stereographs in America and Their Collection* (Gettysburg, PA: Times and News Publishing, 1964), 109–12.
25. Rufus Rockwell Wilson, preface to *Washington through the Stereoscope: A Visit to Our National Capital* (New York: Underwood & Underwood, 1904). I could not find instructions to any of the boxed sets of world's fair stereos that were sold in 1905 but presume that this sort of instruction accompanied them as well.
26. Ibid., 11.
27. Edward Earle, *Points of View: The Stereograph in America—a Cultural History* (Rochester, NY: Visual Studies Workshop, 1979), 111.
28. Underwood and Underwood, *The United States of America through the Stereoscope* (New York: Underwood & Underwood, 1904), 13–14.
29. Ibid., 10–14.
30. Earle, *Points of View*, 16, 95.
31. Ibid., 19.
32. Jonathan Crary, *Techniques of the Observer: On Vision and Modernity in the Nineteenth Century* (Cambridge, MA: MIT Press, 1990), 117.
33. Samuel A. Batzli, "The Visual Voice: Armchair Tourism, Cultural Authority and the Depiction of the United States in Early Twentieth-Century Stereographs" (PhD dissertation, Univer-

sity of Illinois, Urbana, IL, 1997; obtained through University Microfilms, Ann Arbor, MI), 194–95.
34. Beth Rayfield, "Double Dating: Courtship Ritual and the Radical Potential of Stereographic Viewing," *Iconomania: Studies in Visual Culture* (February 1998), 1–7. Laura Schiavo, "From Phantom Image to Perfect Vision: Physiological Optics, Commercial Photography, and the Popularization of the Stereoscope," in *New Media, 1740–1915,* ed. Lisa Gitelman and Geoffrey B. Pingree (Cambridge, MA: MIT Press, 2003), 131. Schiavo is also author of a much longer and more extensive and informative dissertation on stereographs, "A Collection of Endless Extent and Beauty: Stereographs, Vision, Taste and the American Middle Class, 1850–1880" (PhD dissertation, George Washington University, 2003; obtained through University Microfilms, Ann Arbor, MI).
35. Judith Ann Babbitts, "'To See Is to Know': Stereographs Educate Americans about East Asia, 1890–1940" (PhD dissertation, Yale University, New Haven, CT, 1987; obtained through University Microfilms, Ann Arbor, MI), 15, 133, 228.

CHAPTER FIVE

1. Siegfried Kracauer, *Theory of Film: The Redemption of Physical Reality* (New York: Oxford, 1960), x, 16, 19. Kracauer equates film and photography in discussing these basic characteristics.
2. John Tagg writes, "that a photograph can come to stand as *evidence,* for example, rests not on a natural or existential fact, but on a social, semiotic process." *Burden of Representation,* 4.
3. Barthes, *Image-Music-Text,* 28.
4. Lenz Collection, vol. 8. The handwriting identifying Mrs. Wilkins is probably not that of Beals. But she may have written on the back of the photograph, which is not available for viewing. In one photo of a particularly appealing Igorot child known as Singwa, the caption picturing the boy with bow and arrow reads, "A Modern Cupid." Jessie Tarbox Beals, "A Modern Cupid," in photographs, LPE—Exhibits, Anthropology, Philippine Village, image no. 1278–1354, Louisiana Purchase Exposition Photographic Collection, Missouri Historical Society, St. Louis. This photograph has been reproduced several times in documentary books and films, but without comment, as if the authors recognize but do not explore its significance.
5. Margaret Johanson Witherspoon, *Remembering the St. Louis World's Fair* (St. Louis: Folkstone Press, 1973). Eric Breitbart, in *A World on Display,* published this photograph of Mrs. Wilkins without attribution or discussion. He also slightly misquotes the caption. Jane Cutler, in *The Song of the Molimo,* depicts Ota Benga learning to do the cakewalk.
6. Roland Barthes, "Rhetoric of the Image," in *Classic Essays on Photography,* ed. Alan Trachtenberg (New Haven, NY: Leete's Island Books, 1980), 270–74.
7. The only known "publication" of Beals's photos of the Fair is a scrapbook collection held at the Getty Library in Los Angeles. This was apparently an individual collection. It contains a number of Igorot photos and captions, although not that of Mrs. Wilkins.
8. Parezo and Fowler identify the names and ages of Igorot participants at the Fair. Apparently, the youngest male was around thirteen years old. Thus, the "boy" in the picture is at least a teenager. *Anthropology Goes to the Fair,* 415.
9. Laura Wexler, *Tender Violence: Domestic Visions in an Age of U.S. Imperialism* (Chapel Hill: University of North Carolina Press, 2000), 262–90, 5–7.
10. Photographs taken from the reenactment of the Boer War on the fairgrounds joined actual shots taken from the war itself, creating a circulation of images that created a photomontage of the actual and imaginary.
11. Catherine Hodeir, "Decentering the Gaze at the French Colonial Exhibitions," in *Images and Empires: Visuality in Colonial and Postcolonial Africa,* ed. Paul Landau and Deborah D. Kaspin

(Berkeley: University of California Press, 2002), 243–44. She concludes that these villages were designed according to the logistics of the display site and its purposes. Maxwell, *Colonial Photography and Exhibitions*, 3–6, 13, 19–22, 123–27.
12. M. Lerner, "Hoping for a Splendid Summer," 38.
13. Ibid., 33–41. Lerner suggests that both racial segregation and discrimination increased after black migration into the city increased.
14. From the "Minutes of the Committee on Grounds and Buildings," June 9, 1904; and Beverly Bishop, "Analysis of Transcription" Louisiana Purchase Oral History Collection, box 1, file 2, Oral History Transcripts; and commentary by Missouri Historical Society Staff, box 1, file 46: "Negro Day," Louisiana Purchase Exposition, Oral History Transcripts. See also Theo Olsen to Norris Gregg, head of Concessions, July 27, 1904, "Negro Day," box 1, file 46, Louisiana Purchase Oral History Collection.
15. John A. Wright argues that the Fair ultimately inspired a substantial migration of African Americans to St. Louis. See also William Crossland, *Industrial Conditions among Negroes in St. Louis* (St. Louis: Press of Mendle Printing Co., 1914).
16. Michelle Shawn Smith, "Looking at One's Self through the Eyes of Others: W. E. B. Du Bois's Photographs for the 1900 Paris Exhibition," *African American Review* 34 (Winter 2000): 581–99.
17. Bernth Lindfors, "Ethnological Show Business: Footlighting the Dark Continent," in *Freakery: Cultural Spectacles of the Extraordinary Body*, ed. Rosemarie Garland Thomson (New York: New York University Press, 1996), 216–17. In an article for *Atlantic Monthly* in 1885, J. G. Wood recounted several visits to a dime museum in Boston, discussing the various peoples and animals on exhibit. In many respects this popular culture venue resembled the later, more scholarly exhibits at world's fairs in emphasizing the particular physical characteristics of groups and stressing, in some instances, that they were "dying races." J. G. Wood, "Dime Museums: From a Naturalist's Point of View," *Atlantic Monthly* 55 (June 1885): 760–65.
18. Mrs. N. B. Sterrett, "Negro at the Charleston Exposition," *African Methodist Episcopal Church Review* 19, no. 1, 463–66.
19. Jose D. Fermin, *1904 World's Fair: The Filipino Experience* (Diliman, Quezon City: University of the Philippines Press, 204), 40–42.
20. I arrived at this generalization by searching seventeen major American newspapers for articles on the St. Louis World's Fair and the Philippines from 1903–1904. My estimates are based upon 161 articles found during that time period.
21. "Mayor Invites Revolt," *Chicago Daily*, October 24, 1904.
22. Henry Grady, "Philippine Village, Startling Exhibit at Fair," *Atlanta Constitution*, July 3, 1904.
23. *Washington Post*, August 8, 1904.
24. W. J. McGee, "Anthropology," *World's Fair Bulletin* 4 (February 1904): 4.
25. W. J. McGee, "Anthropology at the Louisiana Purchase Exposition," *Science*, 22 (December 22, 1905): 826. This article was derived from the opening chapter of the final report of the Anthropology Department. W. J. McGee, "Report of the Department of Anthropology," box 8, series iii, subseries xi, incomplete record of part 2, LPE Co. MSS.
26. McGee, "Anthropology," *Science*, 815. McGee saw genuine progress in the reduction of racial types from about fifty categories or so of races theorized in the eighteenth century. McGee, addendum to "Report of the Department of Anthropology," box 30, series iii, subseries xi, file 2, 8, LPE Co. MSS.
27. W. J. McGee, "Strange Races of Men," *World's Work* 8 (1904): 5185.
28. James C. Bradford, *Crucible of Empire. The Spanish-American War and Its Aftermath* (Annapolis, MD: Naval Institute Press, 1993), 206–13.
29. McGee, "Anthropology," *Science*, 826. See also Steven Conn, *History's Shadow: Native Americans and*

Historical Consciousness in the Nineteenth Century (Chicago: University of Chicago Press, 2004), on Native Americans, anthropology, and racial theories. McGee, "Strange Races of Men," 5185.
30. McGee was clearly startled by this result. He attributed this "utter lack of athletic ability on the part of the savages to the fact that they have not been shown or educated." McGee, "Report of the Department of Anthropology," 180–86 (quote from p. 186), LPE Co. MSS.
31. McGee, "Strange Races," 5185
32. "The Government Philippines Exposition," *Scientific American* 91 (July 23, 1904): 67. See also "The Racial Exhibit at the St. Louis Fair," *Scientific American* 91 (December 10, 1904): 412, 414. See also Michael L. Krenn, *The Color of Empire: Race and American Foreign Relations* (Washington, DC: Potomac Books, 2006), 49–50; Frank Schumacher, "Small World: Visions of Globalization at the Louisiana Purchase Exposition, 1904," paper delivered to the German Association of American Studies, February 21, 2003 (copy supplied by the author); Rydell, *All the World's a Fair*.
33. A. L. Lawshe et al., *Report of the Philippine Exposition Board to the Louisiana Purchase Exposition* (St. Louis: Greeley Printery, 1904), 48. See Kramer, "Making Concessions." This is the most persuasive account of the relationship of the Fair and notions of an American Empire.
34. Christopher A. Vaughan, "Ogling Igorots: The Politics and Commerce of Exhibiting Cultural Otherness, 1898–1913," in *Freakery*, ed. Thomson, 219–21. Vaughan notes that Filipinos were exhibited at further centennials and exhibitions by the Anthropological Exhibit Company.
35. Parezo and Fowler, *Anthropology Goes to the Fair*, 167. One of the great virtues of this study is to give identity and voice to many of the native peoples imported for the anthropology show.
36. "How a Monkey Got In: Was Mistaken for One of the Igorrote Babies." This story was widely reprinted in the American press: See, for example, "Filipino Savages Arrive," *Cedar Falls Gazette*, April 8, 1904, 10. See also P. V. Bradford and Blume, *Ota*.
37. "Verner Bringing Kaieke and Pigmy," scrapbook, August–November 1906, Scrapbooks of the Fair, box 52, series xiv, LPE Co. MSS. Verner to McGee, February 2, 1904, box 30, file 30, Documents on the Tueki People and the finding of Ota Benga from the S. P. Verner Collection of the Carolina Library of the University of South Carolina, copy in LPE Co. MSS.
38. Leon Litwack, "Hellhounds," in *Without Sanctuary*, ed. James Allen, Hilton Als, Congressman John Lewis, and Leon F Litwack (Santa Fe, NM: Twin Palms, 2003), 8–30.
39. Michael J. Pfeifer, "Missouri Lynchings, 1836–1981," http://academic.evergreen.edu/p/pfeiferm/missouri.html.
40. Ray Stannard Baker, *Following the Color Line: American Negro Citizenship in the Progressive Era* (New York: Doubleday, Page, 1908), 30.
41. Bertram Wilbur Doyle, *The Etiquette of Race Relations in the South: A Study in Social Control* (New York: Schocken, 1971), viii–xv, 143. See also Grace Elizabeth Hale, *Making Whiteness: The Culture of Segregation in the South, 1890–1940* (New York: Pantheon, 1998). Hale makes the astute argument that crossing boundaries between the races could be a form of maintaining them: "Containing the mobility of others allowed whites to put on black face, to play with and project upon darkness, to let whiteness float free," 8.
42. W. E. B. Du Bois, "Of the Coming of John," in *The Souls of Black Folk: Essays and Sketches* (Chicago: McClurg, 1903).
43. For further discussion of these elaborate barriers, see Jane Dailey, Glenda Elizabeth Gilmore, and Bryant Simon, eds., *Jumpin' Jim Crow: Southern Politics from Civil War to Civil Rights* (Princeton, NJ: Princeton University Press, 2000); and Birget Branden Rasmussen, Eric Klineberg, Irene J. Nexica, and Matt Wray, *The Making and Unmaking of Whiteness* (Durham, NC: Duke University Press, 2001).
44. Matthew Frye Jacobson, *Whiteness of a Different Color: European Immigrants and the Alchemy of Race* (Cambridge, MA: Harvard University Press, 1998), 2–9, 33–37, 56, 68.

45. On occasion, some of the village peoples rejected their status as anthropological objects. For example, students belonging to the Visayan group objected to being called "savages." They claimed that they had nothing in common with Moros, Igorots, or Negritos. *Washington Post*, August 21, 1904, 8. Albert Ernest Jenks, *The Bontoc Igorot* (Manila: Bureau of Public Printing, 1905), 5.
46. Leon Guerrero, in *Official Catalogue, Philippine Exhibits*.
47. Thomas F. Gossett, *Race: The History of an Idea in America* (New York: Oxford, 1997), 337–416. See also Matthew Pratt Guterl, *The Color of Race in America, 1900–1940* (Cambridge, MA: Harvard University Press, 2001). We can see the reach of this racial confusion in a story in the *Fresno Morning Republican* a year later. Barbers in the town of Normal, Illinois, refused to shave Filipino students at Illinois Normal (now Illinois Normal State University), as well as blacks, for fear of alienating white customers. *Fresno Morning Republican*, September 27, 1905, 4. The *Arena* magazine published a long article in January 1900 on "The White Man's Problem," which, according to the author, was his inability to understand "why we should add several millions of negroes to our population when we already have eight millions of negroes in the United States." Charles Minor Blackford, Jr., 23, (January 1900): 2. This identification of Filipinos as blacks apparently continued long after the Fair closed. In the 1920s, sociological investigators in Chicago suggested that in dating Filipino men, white women often began a slide into "Black and Tan," see Paul Cressey, "Taxi Dance Halls," in Ernest W. Burgess MSS., box 130, file 2/20/26, Regenstein Library, University of Chicago.
48. Anthony Slide, *American Racist: The Life and Films of Thomas Dixon* (Lexington: University of Kentucky Press, 2004).
49. Thomas Dixon, *The Leopard's Spots: A Romance of the White Man's Burden* (New York: Doubleday, Page, 1902), 242. Thomas Dixon, *The Clansman* (New York: Doubleday, Page, 1905), 290.
50. Du Bois, *The Souls of Black Folk*. The last chapter of this compendium is "The Sorrow Songs," which, in some respects, complements and completes the first essay, "Of Our Spiritual Strivings." A few years after the Fair, Du Bois attended and reported on the First Universal Races Congress, held in London in 1911. This was a one-time-only meeting that actually accomplished little, but Du Bois recognized that the effort to look at the idea and operation of race had immense potential. W. E. B. Du Bois, "The First Universal Races Congress," *Independent* 71 (1911): 401–3.
51. Edward Said, *Orientalism* (New York: Random House, 1994), 190.
52. Jonah Raskin, "Imperialism: Conrad's Heart of Darkness," *Journal of Contemporary History* 2 (April 1967): 131. Raskin is quoting Conrad here.
53. Elia W. Peattie, "On Conrad's Youth and Isham's Under the Rose," *Chicago Tribune* March 23, 1903, 13.
54. Eric J. Sundquist, *To Wake the Nation: Race in the Making of American Literature* (Cambridge, MA: Harvard University Press, 1993), 308. Sundquist argues the larger point that the cakewalk was akin to dialect writing in establishing a "crossroads of cultural languages."
55. Brooke Baldwin, "The Cakewalk: A Study in Stereotype and Reality," *Journal of Social History* 15 (Winter 1981): 205–18. Thomas F. DeFrantz, "Cakewalk," *Greenwood Encyclopedia of African American Folklore* (Westport, CT: Greenwood Press, 2006), 1:184–85.
56. Rachel Blau DuPlessis, "'Hoo, Hoo, Hoo': Some Episodes in the Construction of Modern Whiteness," *American Literature* 67 (December 1995): 688.
57. Edward Coff remembers that the song "Meet Me in St. Louis, Louis" was very popular at the time, as was "Under the Bamboo Tree." Edward J. Coff, "A Boy's-Eye View of the World's Fair," *Missouri Historical Society Bulletin* 11 (October 1954): 71.
58. Jane Anne Liebenguth, "Music at the Louisiana Purchase Exposition," *Bulletin of the Missouri Historical Society* 36 (October 1979): 32.

59. Tatsushi Narita, "The Young T. S. Eliot and Alien Cultures: His Philippines Interactions," *Review of English Studies* 45 (November 1994): 523–24. See also Marjorie J. Lightfoot, "Charting Eliot's Course in Drama," *Educational Theatre Journal* 20 (May 1968): 188.
60. *St. Louis Post-Dispatch*, July 3, 1904, cited in Kramer, *Blood of Government*, 277. There were tamer forms of fraternization that seemed perfectly acceptable because they involved children. Thus, Singwa, the five-year-old "Filipino Prodigy," became a star attraction at the Igorot Village for his verbal skill and artistry. *Republican News* (Hamilton, OH), October 6, 1904.
61. In a very interesting paper, John McKerley outlines the complex attitudes of the black community in St. Louis toward the Fair and its presentations of race and ethnicity. These attitudes, he suggests, largely reflected class divisions in that community, with middle-class members more hostile to identification with Filipinos than others. John McKerley, "An Encounter with Empire: The 1904 St. Louis World's Fair and the Politics of Race in Progressive Era Missouri," paper presented to the annual meeting of the American Studies Association, November 4, 2005, cited by permission of the author.
62. *Chicago Tribune*, July 24, 1904, 2; Rydell, *All the World's a Fair*; *Des Moines Daily Capital*, July 15, 1904. It is fascinating that, in the *Des Moines Daily Capital*, this story was placed next to an item about a white Alabama man who was (to the shock of the paper) fined for "disciplining a niggeah." In early 1905, the Indiana legislature was apparently considering a law that would make marriage between Filipinos and white women illegal. *Ft. Wayne Sentinel*, March 3, 1905, 7; see also *Des Moines Daily Capital*, July 15, 1904.
63. Kevin J. Mumford, *Interzones: Black/White Sex Districts in Chicago and New York in the Early Twentieth Century* (New York: Columbia University Press, 1997). Mumford argues that these zones of interracial contact redefined sexuality. I have extended his concept to define the Fair as a much less overtly sexual and mostly imaginary interzone that nonetheless had the potential to allow mixing. See John D'Emilio and Estelle B. Freedman, *Intimate Matters: A History of Sexuality in America*, 2nd ed. (Chicago: University of Chicago Press, 1997), vi. See also Lisa Duggan, "From Instincts to Politics: Writing the History of Sexuality in the U.S.," *Journal of Sex Research* 37 (February 1990): 95–109.

CHAPTER SIX

1. Clifford Geertz, "Thick Description: Toward an Interpretation of Culture," *The Interpretation of Cultures* (New York: Basic, 1973), 17.
2. "Journal of Mary McKittrick Markham," vol. 1, 1897–1909, from Diaries and Scrapbooks, 1891–1943, Missouri Historical Society, St. Louis.
3. "Laura Merritt's Diary," vol. 1, 1897–1906, from Diaries and Scrapbooks, 1891–1943, Missouri Historical Society, St. Louis.
4. "Diary of Edward Schneiderhahn," in *Indescribably Grand*, ed. Clevenger, 41–50.
5. Edmond Philibert, "World's Fair Diary," in *Indescribably Grand*, ed. Clevenger, 77–109. The letters of Florence McCallion, the wife of a farmer and a visitor to the Fair, reinforce the notion of the idiosyncratic itinerary of visits to exhibits. In *Indescribably Grand*, ed. Clevenger, 121–23.
6. Sam Hyde's diary, quoted in *Indescribably Grand*, ed. Clevenger, 134–43, 133. See also his remarks on the Japanese, where he records his fear of being attacked by "one of those ugly little Japs," 138.
7. "Year of the Fair Great for Small Boy," *St. Louis Post-Dispatch*, May 23, 1954.
8. Emma Helena Kuhn Mohr, interview, box 1, file 20, Oral History Transcripts.
9. Willard B. Shelp, interview transcript and analysis, box 1, file 28, Oral History Transcripts.
10. Randi Korn and Associates, "Summative Evaluation of 'Meet Me at the Fair.'"
11. "Report of the Jefferson Guard: Miscellaneous Remarks," box 16d, series viii, folder 4, 82–83,

LPE Co. MSS. Clayton Laurie, "An Oddity of Empire," *Gateway Heritage* 15, no. 3 (1994–1995): 47–51. The Philippine Scouts represented a further ethnic segregation, consisting mainly of Visayans, Tagalogs, and Bikols.

12. Parezo and Fowler make the unsubstantiated claim that more visitors went to the Philippine Reservation than attended concessions on the Pike or entered the industrial pavilions. Even if it was true, and concession proceeds certainly suggest it was not, there is no way to measure the number of patrons on the Pike. *Anthropology Goes to the Fair*, 192.

13. Department of Concessions, "Accounts as of February 18, 1905," and "Minutes of the Concessions Committee," box 13, series v, subseries i, folder 6, LPE Co. MSS. Minutes of the department indicated that among the concessions not licensed were proposals for an exhibit of the largest and fattest people in the world, a concession to exhibit American skunks, and one to sell moonshine whiskey.

14. Susan J. Matt, "You Can't Go Home Again: Homesickness and Nostalgia in U.S. History," *Journal of American History* 94 (September 2007): 487.

15. William H. A. Williams, *'Twas Only an Irishman's Dream: The Image of Ireland and the Irish in American Popular Song Lyrics, 1800–1920* (Urbana: University of Illinois Press, 1996), 230, 240–42.

16. Matthew Frye Jacobson, *Barbarian Virtues: The United States Encounters Foreign Peoples at Home and Abroad, 1876–1917* (New York: Hill & Wang, 2000), 192.

17. Paul Michael Lützler and graduate students, "The St. Louis World's Fair as a Site of Cultural Transfer: German and German American Participation," in *German Culture in Nineteenth-Century America: Reception, Adaptation, Transformation*, ed. Lynne Tatlock and Matt Erlin (Rochester, NY: Camden House, 2005), 80. Hermann Knauer, *Guide to the German and Tyrolean Alps* (no publication information), 5.

18. Margaret Lo Piccolo Sullivan, *Hyphenism in St. Louis, 1900–1921: The View from the Outside* (New York: Garland, 1990), viii–22, 80. Gary Mormino estimates that in 1880 about two-thirds of the city of St. Louis consisted of German and Irish immigrants or persons of German and Irish origin. Gary Ross Mormino, *Immigrants on the Hill: Italian Americans in St. Louis, 1882–1982* (Urbana: University of Illinois Press, 1986), 17. It is also clear that immigrants from eastern and southern Europe largely avoided St. Louis after the turn of the century.

19. Sullivan, *Hyphenism in St. Louis*, 27, 32. There has been some suggestion that German American sympathy for the Boers might have stimulated the Boer War display. St. Louis businessmen supported the concession even if the British government was uneasy about it. See Ted Hinckley, "When the Boer War Came to St. Louis," *Missouri Historical Review* 61 (April 1967): 286–94.

20. William Barnaby Faherty, S.J., *The St. Louis Irish: An Unmatched Celtic Community* (St. Louis: Missouri Historical Society Press, 2001), 153–54.

21. Thomas R. MacMechen, *The True and Complete Story of the Pike and Its Attractions* (St. Louis: n.p., 1904), 61. Marion R. Casey, "Ireland, New York and the Irish Image in American Popular Culture, 1890–1960" (PhD dissertation, New York University, 1998; obtained from University Microfilms, Ann Arbor, MI), 157, 187–88. She writes, "The image that Irish people residing in the United States consistently sought to put forward for themselves stressed an historically positive relationship with their adopted country as well as the glorious contributions of their homeland to Western civilization," 131. Margaret Sullivan suggests that one element of organized Irish identity, the St. Patrick's Day parade, was at its height in this period. *Hyphenism in St. Louis*, 89.

22. George Lipsitz, *The Sidewalks of St. Louis: Places, People, and Politics in an American City* (Columbia: University of Missouri Press, 1991), 110–11.

23. Milette Shamir, "Ethnography Conference," manuscript, cited by permission of the author. This excellent essay argues that the model of Jerusalem occupied the physical and ideological space

between the technologically progressive sites of the main exhibition and backward-looking concessions of the Pike.
24. "The Pike Parade," *World's Fair Bulletin* 5 (May 1904): 68–69. Based on proceeds, the Cairo exhibit at St. Louis clearly did not command the extraordinary attention it garnered at Chicago.
25. Meg Armstrong, "A Jumble of Foreignness," *Cultural Critique*, Winter 1992, 207–41.
26. Thorstein Veblen, *The Theory of the Leisure Class: An Economic Study of Institutions* (London: Macmillan, 1917), 167, 170, 175.
27. For various contemporary styles, see *Elite Fashion Catalog, 1904, Elite Styles Co.* (Mineola, NY: Dover, 1996).
28. Susan B. Kaiser, *The Social Psychology of Clothing: Symbolic Appearances in Context*, 2nd ed. (New York: Macmillan, 1990), 389.
29. Georg Simmel, "Fashion," *American Journal of Sociology* 62 (May 1957): 546; reprinted from *International Quarterly* 10 (October 1904): 130–55.
30. Ibid., 543, 546.
31. Kurt W. Back, "Modernism and Fashion: A Social Psychological Interpretation"; and Fred Davis, "Clothing and Fashion as Communication," both in *The Psychology of Fashion*, ed. Michael R. Solomon (Lexington, MA: Lexington Books, 1985), 6, 11, 20–25.
32. "Report of the Jefferson Guard," 87. Similarly, while the naked bodies of Filipinos were eroticized, the audience probably did not notice the large number of nude statues and paintings in the fine arts collection and outdoor statuary. See, for example, the U.S. Government Building which featured a goddess of liberty in a chariot pulled by four horses and guided by two nude male figures.
33. *Report of the Commission: Kentucky at the World's Fair: St. Louis, 1904* (Report to the Governor, 1905), 57; and "Dedicate Filipino Exhibits at the Fair," no citation, from "Scrapbooks of the Fair," August–November 1906, box 52, series xiv, LPE Co. MSS.
34. Board of Lady Managers, *Report to the Louisiana Purchase Exposition Commission* (Cambridge, MA: Riverside Press, 1905), 32–34.
35. Vaughan, "Ogling Igorots," 224–25. *World's Fair Bulletin,* September 1904, 48.
36. Frederick Starr, quoted in *Decatur Review,* August 26, 1904. "Anthony Comstock Sees Naked Igorots," *Ogden Standard,* July 7, 1904. Comstock could not close the Chicago performances, but he did succeed in shutting down the *danse du ventre* when it opened in New York City in 1894.
37. Albert Ernest Jenks, "Bontoc Igorot Clothing," *American Anthropologist*, n.s. 6 (October–December 1904): 700–704. See also Mary Ellen Rouch and Joanne Bubolz Eicher, "The Language of Personal Adornment," in *The Fabrics of Culture: The Anthropology of Clothing and Adornment*, ed. Justine M. Cordwell and Ronald A. Schwarz (The Hague: Mouton, 1979), 20. The authors suggest that in tropical countries where people remain naked in public, it is ornament that signals sexuality.
38. For a discussion of the complexities of nudity and clothing, see Ruth Barcan, *Nudity: A Cultural Anatomy* (Oxford: Berg, 2004), 151–62. Valerie Steele writes that "at the deepest level all clothing is erotic." *Fashion and Eroticism: Ideals of Feminine Beauty from the Victorian Era to the Jazz Age* (New York: Oxford University Press, 1985), 9.
39. "Clothing the Savages," *Washington Post,* July 17, 1904. There is a hint in the tone of this article that Antonio is being used as an ironic mouthpiece for the author.
40. Puck, "The Igorrote's Burden," *Washington Post,* October 2, 1904.
41. Jose D. Fermin, *1904 World's Fair: The Filipino Experience* (Diliman, Quezon City: University of the Philippines Press, 2004). See also, Robert C. Galloway, "Rediscovering the 1904 World's Fair: Human Bites Human," May 2001, www.webster.edu/~corbetre/dogtown/fair/galloway.html. Galloway's angry but interesting essay rejects "cultural relativism" and defends the bringing of Igorots for display.

42. Marlon E. Fuentes, *Bontoc Eulogy* (Independent Television Service and National Asian American Telecommunications Association, 1995), documentary.
43. See also Pennee Bender et al., *Savage Acts: War, Fairs, and Empire* (American Social History Productions, 1995), documentary. This short video traces the simultaneous history of American imperial adventures in the Philippines and the depiction of savage peoples at world's fairs.
44. Dorothee Wierling, "The History of Everyday Life and Gender Relations: On Historical and Historiographical Relationships," in *The History of Everyday Life: Reconstructing Historical Experiences and Ways of Life*, ed. Alf Ludtke, trans. William Templer (Princeton, NJ: Princeton University Press, 1995), 151.
45. Department of Agriculture and Horticulture, "Attendance in Building," box 7a, folder 1, 69, LPE Co. MSS.
46. "Farmers' Month at World's Fair," *Semi-Weekly Waterloo Courier*, August 30, 1904.
47. Interview with Mrs. Emile Waringer, box 1, file 35, and interview with Willard Shelp, box 1, file 28, 12, Oral History Transcripts.
48. Some of the state pavilions still exist as residences.
49. *Report of the Maryland Commission to the Louisiana Purchase Exposition to the General Assembly of Maryland*, session 1906 (Baltimore, MD, 1906), 32–34, 105.
50. Delancy M. Ellis, *New York at the Louisiana Purchase Exposition: Report of the State Commission* (Albany, NY: J. B. Lyon, 1907), 19, 67–68.
51. James H. Lambert, *The Story of Pennsylvania at the World's Fair, St. Louis, 1904*, 2 vols. (Philadelphia: Pennsylvania Commission, 1905), 1:263.
52. President Cleveland quoted in Ellis, *New York at the Louisiana Purchase Exposition*. He is probably quoted in this book because he was a New Yorker. "The St. Louis Fair," *New York Times*, May 1, 1904. Lambert, *Story of Pennsylvania*, 2:38–39. Although it is likely that close to 25,000 different individuals entered the Pennsylvania building on Independence Day, the summer-long figures no doubt included many repeat visits.
53. [Henry] Rolstair's Creation: The Formation of the Earth and Its Inhabitants on the Pike was popular among patrons and contradicted the predominant evolutionary schema that organized most of the Fair.
54. Everett, *Book of the Fair*, 148.
55. Mark Bennett, *History of the Louisiana Purchase Exposition* (St. Louis: Universal Exposition Publishing Co., 1905), 742.
56. Skiff, preface to *Official Catalogue of Exhibitors*.
57. Edward Schneiderhahn, "Diary," vol. 7, May 19, 1904, Journals and Diaries, Missouri Historical Society.
58. *World's Fair Authentic Guide*, 157.
59. "Daily Official Program: June 2," Louisiana Purchase Exposition, LPE Co. MSS. See also *World's Fair Bulletin: Official List of Concessionaires*, April 1904; and "Conventions, Congresses and Special Meetings Held in St. Louis during 1904," box 11, series iv, subseries ii, folder 1, LPE Co. MSS.
60. David Francis, June 4, 1904, David Rowland Francis Papers, Missouri Historical Society, St. Louis.
61. Department of International Congress, "List of International Congresses, Giving Meeting Place and Attendance of Each," 1904, box 9, series 3, subseries 15, file 1, 114–23, LPE Co. MSS. I added their figures together. These are rounded-off estimates and purport to represent different individuals.
62. See the interesting essay by Alexander Moore on Walt Disney World as a pilgrimage site. Disney, it should be said, reproduced many of the elements of world's fairs in his amusement

parks. Moore, "Walt Disney World: Bounded Ritual Space and the Playful Pilgrimage Center," *Anthropological Quarterly* 53 (October 1980): 207–17.
63. Amatai Etzioni and Jared Bloom, *We Are What We Celebrate: Understanding Holidays and Rituals* (New York: New York University Press, 2004), 10.
64. Using the online source NewspaperArchive.com, I searched for "family reunion" in selected years in the 1880s and then every year after 1890 until 1916. The number of stories concerning family reunions was modest in the beginning (less than 100), over 1,000 by 1900, and almost 2,000 five years later. There are reasons not to rely on these figures as precise, but the general trend is clear. See also Robert M. Taylor, Jr., "Summoning the Wandering Tribes: Genealogy and Family Reunions in American History," *Journal of Social History* 16 (Winter 1982): 21–37.
65. John R. Gillis, *A World of Their Own Making: Myth, Ritual, and the Quest for Family Values* (New York: Basic, 1996), 5–6.
66. Etzioni and Bloom, *We Are What We Celebrate*, 45.
67. Gwen Kennedy Neville, *Kinship and Pilgrimages: Rituals of Reunion in American Protestant Culture* (New York: Oxford, 1987), 23–26.
68. Susan Matt, "You Can't Go Home Again," 479, 490–91.
69. E. C. Culp, "Incidental Information," minutes of the Committee on Ceremonies, March 3, 1904, copy in box 1, file 46, Oral History Transcripts.
70. C. Howard Walker, "The Louisiana Purchase Exposition at St. Louis, Missouri," *Architectural Review*, August 1904, 201.
71. Department of Concessions, "Report," 1904, box 13, series v, subseries i, folder 1a, 7–9, LPE Co. MSS. The department estimated that unauthorized sales at foreign exhibits amounted to another $1,000,000.
72. Jane P. Tompkins, ed., *Reader-Response Criticism: From Formalism to Post-Structuralism* (Baltimore: Johns Hopkins University Press, 1980), x–xxiii.

CHAPTER SEVEN

1. "Negro Day," file 46, 199–202, reproduced from Oral History Transcripts.
2. "Propose Bureau for Negro Race," *St. Louis Globe-Democrat*, June 2, 1904, 3; W. M. Farmer to Booker T. Washington, April 8, 1904, Booker T. Washington Papers, container 802, Library of Congress.
3. W. M. Farmer to Booker T. Washington, July 28, 1904, Booker T. Washington Papers, container 288, Library of Congress.
4. "Colored Women Stay Away from Fair," *St. Louis Globe-Democrat*, July 13, 1904. Even if there was job discrimination, the *St. Louis Palladium* (a black newspaper) published several articles about the Fair, including a discussion of the many new jobs available. *World's Fair Bulletin* (1904 World's Fair Society) 7 (October 1992).
5. Gerda Lerner, "Early Community Work of Black Club Women," *Journal of Negro History* 59 (April 1974): 158–67.
6. J. S. Yates to Margaret Murray Washington, January 15, 1902, in *National Association of Colored Women's Clubs Papers, 1895–1992*, ed. Lillian Serece Williams, part 1, reel 5, "Minutes of National Conventions, Publications and President's Office Correspondence," in *Black Studies Research Sources: Microfilms of Major Archival and Manuscript Collections*, ed. John H. Bracey, Jr., and August Meier. Hereafter cited as NACWC MSS.
7. "Minutes of the National Conventions," part 1, reel 5, NACWC MSS.
8. "Minutes of the Executive Committee," July 15, 1904, and "Minutes of the Evening Session," *Minutes of the Third Biennial Meeting or Fourth Convention of the National Association of Colored*

Women, St. Louis, Mo., July 11–16, 1904 (Jefferson City: Hugh Sterns, [1904?]), 10, July 13, 1904, reproduced in part 1, reel 1, NACWC MSS.

9. "Minutes of the Afternoon Session," July 12, 1904, in part 1, reel 1, NACWC MSS. "Minutes," evening session, July 13, 1904, *Minutes of the Third Biennial Meeting*, 10, in part 1, reel 1, NACWC MSS.

10. Jacobson, *Barbarian Virtues*, 217, 264. Jacobson asserts that the Philippine War and consequent long-term occupation of the archipelago is a central but often-overlooked element of American history.

11. President Theodore Roosevelt used these terms in his Romanes Lecture in 1910 at Oxford. Speaking of "Biological Analogies in History," he employed these terms but in their much more common contemporary usage in which *barbarism* and *savagery* and *civilized* and *enlightened* were interchangeable synonyms.

12. Audrey L. Olson, *St. Louis German, 1850–1920: The Nature of an Immigrant Community and Its Relation to the Assimilation Process* (New York: Arno Press, 1980), 165–67. Olson notes the abrupt decline of German identity, even around *gemutlichkeit* during World War I. See also *Sioux Valley News*, October 13, 1904.

13. John Fiske, *Understanding Popular Culture* (London: Routledge, 1990), 25. Fiske argues that people resist and transform prescribed culture.

14. The Old Plantation was also the name of a national minstrel touring company.

15. Speaking of the *danse du ventre* at the Chicago World's Fair, the *New York Olean Democrat* perceptively noted: "This is, we believe, the first instance in which a college professor [Frederic W. Putnam] has superintended so remarkable an addition to our amusements," September 26, 1893.

16. One of the best general accounts of the Louisiana Purchase Exposition and a complex exploration of the visitor's reaction to it is Martin Brueggemann, "St. Louis' 1904 World's Fair."

17. Michel de Certeau, *The Practice of Everyday Life* (Berkeley: University of California Press, 1984), xviii.

18. Philippine Exposition Board, *Report of the Philippine Exposition Board to the Louisiana Purchase Exposition and Official List of Awards . . . , World's Fair, St. Louis, Mo.* (St. Louis: Greeley Printing, 1904), 40. Abraham L. Lawshe was chair of the board appointed by the War Department.

19. James McDonald, "The City, the Cinema, Modern Spaces," in *Visual Culture*, ed. Chris Jenks (London: Routledge, 1995), 79.

INDEX

Page numbers in italics denote illustrations.

Adams, Henry, 55, 65
African Americans, 130–34, 139–52, 185–89, 192, 207n13, 207n15. *See also* Louisiana Purchase Centennial Exposition: race and racism at; St. Louis: ethnicity and race in
African Pygmies, 39, 60–61, 109, 136, 140
Ainu (Japanese), 39, 136
audience, and history, 3, 7. *See also* experience; visitor experience of the Fair

Baker, Ray Stannard, 141
Barthes, Roland, 105, 124, 126
Batzli, Samuel, 117
Beals, Jessie Tarbox, 96, 102, 110, *110*, 124, 126–27, 130, 143, 206n7
Beaux-Arts design, 18, 65–66
Benedict, Burton, 56, 95
Benga, Ota, 60–61, 93, 96, 140
Bennett, Mark, 53–54, 177
Benson, Sally, 5, 87–88, 94, 98
Blackmer, Paul, 15–16, 35, 192
Bloom, Jared, 181
Boas, Franz, 58, 193
Bodnar, John, 74
Boer War, 116, 120, 131, 157; and concession on the Pike, 161; reenactments of, 43, 61, 66, 108, 158, 191, 206n10; veterans of, 156
Breitbart, Eric, 64, 69–70, 129
Burke, Peter, 105
Busch, Adolphus, 164

cakewalk, 87, 124–25, 130, 146–50, *147*, *148*, *150*
cannibalism, 39–40, 45, 193
Celik, Zeynep, 70
Chicago Century of Progress, 92
Chicago World's Fair. *See* World's Columbian Exposition (Chicago)
China, 38, 118, 166

City Beautiful movement, 29, 65
Cleveland, Grover, 175
Clevenger, Martha, 99
Comstock, Anthony, 169
Conrad, Joseph, 145
consumer culture, 17, 65, 67
Coombes, Diane, 29
Corbett, Katharine, 95–97
culture wars, 74, 201n9

Daughters of the American Revolution, 21, 181
de Certeau, Michel, 193–94
Disneyland, 17, 164, 213n62
Dixon, Thomas, 143–44
dog-eating, 40, 44–46, 93, 100, 108, 111, 113, 135, *140*, 193
Douglass, Walter B., 85
dress, 166–71, 212n37
Du Bois, W. E. B., 27, 66, 132–33, 142, 144, 188

Enola Gay exhibition, 75
Eskimos, 100, 156
Etzioni, Amatai, 181
Everett, Marshall, 39–40, 176
experience, 3–4, 6, 7, 11, 14, 60, 67, 152–53, 160–62, 167–69, 172–84, 189–90, 193; and auditory limitations to, 8, 21; definition of, 7; and history, 14, 53, 60, 79, 153–54, 189–90, 192; and memory, 14, 53, 71, 79, 98, 153, 158; pitfalls of ignoring, 8; of youths versus adults at the Fair, 79, 100, 174

family reunions, 101, 157, 173, 181–82, 185, 188
Farmer, W. M., 186
fashion. *See* dress
Filipinos, 11, 40, 43, 45–46, 50, 60, 62, 66, 87, 100, 111, 113, 124–25, *127*, 132, *133*, 135–37, 139–40, 143, 145–46, 150–51, 157, 160, 167–72; and the American color line, 143, 149–50, 209n47, 210n63. *See also* Philippine Constabulary

Filipinos (*continued*)
  Brass Band; Philippine ethnic groups; Philippine Reservation; Philippines; Philippine school; Philippine Scouts; Philippine War; United States Philippine Commission
Flower, Don, 62
France, 13, 27, 58, 66
Francis, David, vii, 9–10, 18, 20, 22–23, 26, 31, 34, 37–39, 45, 48, 52, 58, 65, 83–84, 86, 94, 96, 107, 132, *140*, 164, 175, 180, 185–86
Frisch, Michael, 74–75
Fuentes, Marlon, 172

Garland, Judy, 5, 88–89, 94, 98, 101, 149–50, *150*
Gateway Heritage, 92
Geertz, Clifford, 154
Gerhard, Emme and Mamie, 111, 131
German Americans, 23, 26, 41, 47, 50, 80–81, 88, 91, 94–95, 142, 156–57, 159–65, 174, 179–80, 191–92
Germany, ix, 13, 27, 32, 43, 47, 58, 81
Geronimo, 30, 131, 157, 192
Gillis, John, 73
Glassberg, David, 20, 95
Grady, Henry, 135
Great Britain, 58
Grindstaff, Beverly, 60

Halbwachs, Maurice, 73
Harris, Neil, 57, 70–71, 95
historians: and archival sources, 37, 52–54; discounting audience experience, 3, 67; evaluating experience, 7, 53, 55, 79, 99, 153–54, 156, 189, 192; evaluating memory, 55, 67, 69–71, 75–76, 78, 99, 101, 189, 195n1 (intro.); and oral history sources, 10, 69–72, 76–77, 79–80, 90, 99–101; and photographic sources, 11, 105, 123–24
historiography, 2; historiographical consensus regarding world's fairs, 64–65; of stereographs, 117–18; of the St. Louis World's Fair, 9, 53–66, 199n38, 200n42, 200n52; and of the study of memory, 72–77
history: and experience, 7, 14, 53, 60, 79, 153–54; and memory, 1–2, 4, 55, 59, 67–68, 70–72, 75, 93, 97, 101, 189, 201n4
Hobsbawm, Eric, 73

Holley, Marietta, 49, 51
Holocaust memory, 72–73, 75–76, 201n11
Hrdlicka, Ales, 58
Hyde, Sam, 156

Indian Reservation, 65. *See also* Native Americans
Irish Americans, 23, 91, 94–95, 142, 147, 157, 159–65, 174, 180, 211n18, 211n21. *See also* St. Louis: ethnicity and race in

Jackson, C. S., 107–8
Jacobson, Matthew Frye, 163, 189
Japan, 118, 142, 166
Japanese, 40, 43, 81, 155, 157, 166
Jefferson Guard, 160, 168, 175
Jefferson Monument, 84, 86
Jenks, Albert Ernest, 28, 142, 170
Jerusalem, 24, 33, 43, 47, 85, 108, 165
Johnston, Frances Benjamin, 111–12
Jojoa, Ted, 70
Joplin, Scott, 63, 96, 158

Kracauer, Siegfried, 123
Kramer, Paul, 62

Landsberg, Alison, 76
Le Goff, Jaques, 72–73
Lipsitz, George, 76
Louisiana Purchase, 137
Louisiana Purchase Centennial Exposition: Agriculture Building, 155, 173–74, 178; anthropological exhibition methods of, 28–32, 39, 63, 135–37, 143; anthropology at, 31, 42, 45, 56–57, 60, 63, 136–38, 142, 160, 200n47; art at, 42, 81, 85; attendance figures of, 15–16, 195nn1–2 (chap. 1), 195n6; baby incubator, 67, 192; cataloguing system of exhibits, 178–79; Congress of Arts and Sciences, 9, 26; as convention center, 179; Creation display, 156, 161, *170*, 213n53; Department of Physical Culture, 137; and economic development, 33–34, 159; education at, 24, 38, 41–42, 47, 107, 135, 200n47; eroticism at, 31, 35, 100, 145, 152, 158, 165, 169–71, 210n63; ethnology at, 29–30, 63, 111, 136, 160, 200n47; and European culture, 143; evolutionary schema of, 26, 28, 43, 64, 86, 137, 143, 160,

· 218 · INDEX

193; exhibition halls, 172–79; exhibits, 25–26, 175–79 ; exoticism at, 32–33, 46, 59, 99–100, 106, 111, 126, 139, 142, 144, 166 (*see also* Orientalism); Ferris wheel, 67, 160; finances, 22, 34; German Beer Garden and town, 91, 161–65; guidebooks, 9, 46–49, 83, 106, 176, 179; Hereafter (concession), 100, 161; histories of, 38–40, 53–56, 83, 177, 199n38, 200n42, 200n52; and identity, 162–64, 171, 174–84, 185, 188, 191, 194; and imperialism, 138, 172, 189–91, 193, 197n40, 213n43; intellectuals' participation, 26–27, 48, 63; Irish Village, 47, 91, 161, 164–65, *168*, 174; Liberal Arts Building, 81; location, 24; machinery exhibition, 42; Manufactures Building, 48, 81; and modernism, 63 (*see also* modernity); Negro Day (Emancipation Day), 132, 159, 186; newspaper and magazine coverage, 41, 42–46, 61, 80, 82, 192; novels about, 9, 46, 49; official catalogue, 28; official publications, 41; one hundred-year anniversary celebration, 93–94; oral histories relating to, x, 10, 90, 100–1, 154, 157–60, 173–74, 176, 179; Palace of Electricity, *107*; Palace of Fine Arts, 85, 143; Palace of Machinery, *177*; Palace of Varied Industries, *107*; personal accounts of, 10, 80–82, 154–57, 179; photographs of, 84, 102–4, 106–12 (*see also* photography); the Pike, 8, 11, 16, 22, 24, 30, 33, 41, 45, 47, 59, 63, 80–82, 91, 106–7, 112, 120, 145, 152, 158, 160–61, 164–70, *168*, *170*, 172, 191–92; popularity of particular concessions, 161–62, 183; proximity, displays and exhibits of, 174–75; publicity, 9, 82; race and racism at, 27–32, 38–41, 42–47, 58, 60, 62–63, 83, 91, 109, 130–52, 156, 159–60, 185–89, 192; seventy-fifth anniversary celebration of, 90; souvenir photo books, 106–8, 120; state buildings at, 44, 172, 174; Temple of Fraternity, 180; Transportation Building, 81; travel to, 24; treatment of women at, 38, 83; Tyrolean Alps, 47, 156, 161–64, *162*, 174, 192; United States Government Building, 81, 155; Varied Industries Building, 81; zoo, 18, 91, 108
Louisiana Purchase Centennial Exposition Company, 9, 66, 82, 84–86, 108, 129, 132, 154, 186; Department of Concessions, 35, 183; Division of Exploitation, 40; Division of Press and Publicity, 9, 16, 41; Executive Committee, 20, 22, 86,
Louisiana Purchase Historical Association, 10, 38, 84, 86
Lowenthal, David, 76
lynching, 141, 187

Markham, Mary McKittrick, 154
Maxwell, Anne, 131
McAllister, Melani, 165
McGee, William J., 30–31, 39, 43, 58, 61–62, 85–87, 99, 134–38, 140, 142–43, 155, 160–61, 172, 193
Mead, Margaret, 193
Meet Me in St. Louis, 5, 87–89, 149–50, *150*; and the collective memory of the Fair, 6, 92, 98, 101; final scene at the Fair, 202n36
"Meet Me in St. Louis, Louis," 71, 182
memory: construction of, 46, 72–78, 83, 95, 97, 118; and cultural literacy, 74, 76; and experience, 14, 76, 79, 82, 98, 156–58; and history, 1–2, 4, 55, 59, 67–72, 75, 93, 97, 101, 172, 189, 201n4; and identity, 73, 159, 164–65, 174, 189; individual versus collective, 7, 68, 78–81, 98, 101, 158–59, 172; institutions of, 83–86, 89; language of, 68; mass media's influence on, 67, 73, 76, 82–83; objects, 92, 97, 204n68; politics of, 74–75, 77, 97, 172, 201n9; problems of, 2–3, 6, 69–70, 80–81, 201n16; and a usable past, 98; vernacular, 74. *See also* historians: evaluating memory; history: and memory; Holocaust memory; memory culture, four divisions of
memory culture, four divisions of, 71–72. *See also* St. Louis: civic identity and memory culture
Merritt, Laura, 155–56
methodology, 4, 14
Miller, Howard S., 95
Missouri Historical Society, 79, 85–86, 89, 90–91, 94, 99, 120, 158–59; "Meet Me at the Fair" exhibition, 94–97, 159, 204n61
Missouri State Federation of Women's Clubs, 187
Misztal, Barbara, 73
modernity, 63, 65–66, 72, 78, 152, 192, 194
Mohr, Emma Helena Kuhn, 157

INDEX · 219 ·

Mumford, Kevin, 151
Munsterberg, Hugo, 26–27, 66, 109, 189

National Association of Colored Women's Clubs, 185–88
National Creamery Men, 48
nationalism, 34, 66, 172, 180–81, 188, 191
Native Americans, 18, 29, 39–40, 45, 61, 63, 65–66, 107, 117, 120, *125*, 131, 136–37, 157, 192
Newell, Alfred C., 40
New York World's Fair, 15, 92
1904 World's Fair Society, 92, 97
Nora, Pierre, 71
nostalgia, 5–6, 29, 68, 71, 89, 101, 159, 162–63, 172, 181–82, 189
nudity, 62, 126–27, 134, 139, 146, 150, 152, 169, 171, *179*, 212n32, 212n37, 212n38

O'Brien, Margaret, 5, 88, 149
Orientalism, 144–45, 165–66

pageantry, 19–22
parades, 22, 120–21, *121*, 156, 175, 191
Parezo, Nancy, 31, 61–62
Paris fairs, 14, 29
Patagonians, 39, 136, 156
patriotism, 20–21, 33–34, 74, 175–76, 181–82, 187
Philibert, Edmond, 156
Philippine Constabulary Brass Band, 155, 168, 175
Philippine ethnic groups: Bagobo, 87, 139; Igorot, 40–41, 44–45, 50, 62, 69, 79, 87, 90, 93, 97, 99–100, 108, 111, 113, 124–30, *126*, *127*, *128*, *129*, *133*, 134–35, 138–39, *140*, 144–45, 147, 149, 152, 156–58, 160, 169–72; Moro, 40, 87, 139; Negrito, 58, 111–*12*, 139; Visayan, 139, 160, 209n45
Philippine Reservation, 28, 32, 39–40, 42–45, 47, 56, 58, 62–63, 81, 93, 99, 109, 112, 120, 134–35, 137–39, 142, 151, 160–61, 169, 172, 183, 193; and riot at Cafe Luzon, 150–51
Philippines, 96, 142, 189, 193
Philippine school, 44–45
Philippine Scouts, 33, 41, 45, 62, 150–51, 160, 168, 175, 199n19
Philippine War, 120, 137, 215n10
photography, 11, 49, 64, 102–22, 192; amateur, 111, 113; and memory, 102, 120; and official views of the Fair, 106, 113, 122; on postcards, 112–13; and types of photographs from the Fair, 104, 124–27; and the use of photographs as sources, 11, 16, 64, 80, 102, 104, 123–24
Primm, James Neal, 55
Putnam, F. W., 30–31

ragtime, 63, 66, 91, 131, 146
Rau, William H., 108–9
Ricoeur, Paul, 71, 77
Roosevelt, Theodore, 8, 109, 135, 172, 196n24
Roosevelt, Alice, 45, 154
Roots, 76
Rozensweig, Roy, 75, 78
Ryan, Mary, 19
Rydell, Robert, 4, 54, 57–60, 70, 95, 200n42

Said, Edward, 144–45
Schiavo, Laura, 118, 206n34
Schneiderhahn, Edward V. P., 80–82, 155–56, 179
Shelp, Willard B., 100, 158, 174
Simmel, George, 167
Singwa (Igorot child), 130, 206n4, 210n60
Skiff, Frederick J. V., 25
Sontag, Susan, 105
Sousa, John Philip, 34, 147, 158
Spanish American War, 43, 44, 57, 65–66, 117–18, 120, 130
Steffens, Lincoln, 20
stereographs, 102, 105, 113–22, 192, 206n34; images from, *114*, *133*, *163*, *178*; and imperialism, 118, 120, 131; and memory, 118; versus photographs, 122–23; and tourism, 116–17
Stevens, George M., 49–51
Stevens, Walter, 54, 108–9
St. Louis: Agricultural and Mechanical Fair, 18; Carnival Week, 19; civic identity and memory culture, 4, 10, 56, 65, 89–90, 92–95, 97–98, 159, 171, 189, 191, 193–94, 211n21; civic importance of *Meet Me in St. Louis*, 6; Democratic National Convention, 41, 56; ethnicity and race in, 23, 95, 141, 159–65, 180, 191, 196n22, 207n13, 207n15, 211n18, 215n12; National Education Association Annual Meeting, 41; Olympic Games, 12, 56, 101, 137

St. Louis Globe-Democrat, 89
St. Louis Post-Dispatch, 79, 93, 157
St. Louis Public Museum, 85
St. Louis—That Fabulous Summer (film by Charles Guggenheim), 90
St. Louis World's Fair. *See* Louisiana Purchase Centennial Exposition
Sundquist, Eric, 146
Susman, Warren, 67–68

Taft, William, 62, 172
Thelen, David, 75, 78, 95
Trouillot, Michel-Rolph, 86
Troutman, John, 31, 61

"Under the Bamboo Tree," 147–49
Underwood and Underwood, 115–17, 119
United Daughters of the Confederacy, 21, 180
United States Philippine Commission, 62, 160

Veblen, Thorstein, 166
Veiled Prophet Pageant Society of St. Louis, 18, 20, 196n22, 197n26
Verner, S. P., 140
Vietnam Veterans Memorial, 74
visitor experience of the Fair, x, 3–4, 7, 9, 11, 25, 31, 36, 43–50, 53, 56, 58, 60, 62, 64, 67, 69–71, 79, 82–83, 98–100, 102, 118, 123, *133*, 134–35, 138–39, 153–57, 160–61, *168*–69, 173–76, 179, 183, 189–93; child visitors, 16, 79–80, 86, 200n52. *See also* experience
Voices, 92

Wagner, Theodore P., 157
Washington, Booker T., 132, 186
Washington, Mrs. Booker T. (née Margaret Murray), 186
Washington Post, 171
Washington University, 24, 26
Wells, Ida B., 187
Wexler, Laura, 130
Whiting View Company, 119
whose fair, x, 6, 190
Wilkins, Mrs., 124–30, *127–29*, 138–39, 141–47, 149–53, 166, 192, 206n4
Williams, William, 162
Wilson, Woodrow, 193
Witherspoon, Margaret Johnson, 125
Wizard of Oz, The, 89
Women's Christian Temperance Union, 48
World on Display, A, (Brietbart), 69–70
World's Columbian Exposition (Chicago), 4–5, 14, 32, 35, 66, 106, 112, 169
World's Fair Bulletin, 92
world's fairs, 4–5, 13–14, 17–18, 23, 186–87; attendance figures of, 8, 14–16; historical interpretation of, 14; in relation to contemporary entertainments, 19–20, 41, 57, 207n17; and tourism, 19–20, 24, 44